When The Scientist Presents

An Audio and Video Guide to Science Talks

Jean-Luc Lebrun
Trainer of researchers and scientists from A*STAR Research Institutes, Singapore
Former Director, Apple-ISS Research Centre, Singapore

NEW JERSEY · LONDON · SINGAPORE · BEIJING · SHANGHAI · HONG KONG · TAIPEI · CHENNAI

Published by

World Scientific Publishing Co. Pte. Ltd.
5 Toh Tuck Link, Singapore 596224
USA office: 27 Warren Street, Suite 401-402, Hackensack, NJ 07601
UK office: 57 Shelton Street, Covent Garden, London WC2H 9HE

Library of Congress Cataloging-in-Publication Data
Lebrun, Jean-Luc.
   When the scientist presents : an audio and video guide to science talks / Jean-Luc Lebrun.
    p. cm.
   Includes index.
   ISBN 9789812839206 (pbk.)
   ISBN 9812839208 (pbk.)
   ISBN 9789812839190 (hardcover)
   ISBN 9812839194 (hardcover)
   1. Communication in science. 2. Public speaking. I. Title.
   Q224 .L43 2010
                       2010278527

British Library Cataloguing-in-Publication Data
A catalogue record for this book is available from the British Library.

First published 2010
Reprinted 2013, 2015

Copyright © 2010 by World Scientific Publishing Co. Pte. Ltd.

All rights reserved. This book, or parts thereof, may not be reproduced in any form or by any means, electronic or mechanical, including photocopying, recording or any information storage and retrieval system now known or to be invented, without written permission from the Publisher.

For photocopying of material in this volume, please pay a copying fee through the Copyright Clearance Center, Inc., 222 Rosewood Drive, Danvers, MA 01923, USA. In this case permission to photocopy is not required from the publisher.

Typeset by Stallion Press
Email: enquiries@stallionpress.com

Printed in Singapore

PREFACE

Vladimir's backstage exit

Vladimir brings the last slide on the screen, a bright white "Q&A" on a dark background. Finally moving his eyes away from the computer screen, he says, "Thank you for your attention. Any questions?"

About ten to fifteen people are scattered in a room too big for an audience this small. Most of them are sitting either in an aisle seat or in the last row, ready to make a quick exit. Some already have. As he waits, he feels quite uncomfortable. Not one hand goes up. In the front row, his supervisor is packing up, ready to leave. After an unbearable four-second silence, Vladimir looks sideways to the chairperson who, after glancing furtively at his watch, rescues Vlad with a loud "Thank you Dr. Toldoff". Vladimir presses the spacebar one last time. The screen goes dark, Microsoft PowerPoint exits front stage, and Vladimir exits backstage.

Outside the conference room, during the coffee break, Vladimir approaches his supervisor, a burning question on his mind.

"How did I do?" he asks.

"You did fine", responds his supervisor while avoiding eye contact.

To get an honest answer, Vladimir changes his question.

"Did I do as well as Dr. Sorpong?"

The supervisor now looks at him, condescendingly.

"As well as Dr. Sorpong? Not quite. I think it would be good if I added presentation skills as an objective in your upcoming performance review. Quite frankly, you need more confidence, more dynamism, better slides, and you need to work on that Russian accent of yours. I am used to it, but the audience clearly isn't."

Vladimir sighs.

"Are you sending me to a presentation skills class?"

"That could be arranged."

"In Hawaii?"

"Don't push it, Vlad."

Greetings! Your name may not be Vladimir, and you may not be Russian, but you are a scientist, and that tells me a lot about you. You are someone who is not satisfied with status quo situations. You suspect there is no such thing as a presentation skills gene in anyone's DNA and you are convinced that these skills are acquired and improved through learning and doing. You have seen great presenters tackling the most difficult topics without effort and with great result. If they can do it, so can you! Great!

Your future skills are sleeping, waiting to be brought to life in this book and the companion DVD. They are here for you, ready to serve you, ready to turn you into a great presenter. How do you access them? Just rub your mind against the words of these pages, and review the media on the DVD. These skills can become yours.

 Where there are exercises, do them; where there are checklists, follow them;

 where there are videos, watch them;

 where there are podcasts, listen to them;

 and where there are websites, visit them.

But most importantly, apply these newfound skills as early as possible.

In this book, we will imagine that your task is to present your paper at a scientific meeting. If all goes well, your presentation will end up being quite different from your paper. It cannot be otherwise because the paper is no longer essential this time, but you are. Danish professor Peter Sigmund of Odessa University, a man who must have seen many bad presentations, wrote a tongue-in-cheek column in Physics Today entitled "Fifteen Ways to Get Your Audience **to Leave You**". The eighth point of advice recommends that you "*ignore the inherent difference that exists between oral and written communication*". Since you want your audience **to stay**, you want to know the difference. That is your first task. For this task, consult the first four book chapters as well as the DVD. They reveal how to select the contents for your slides and prepare you for the slide design and creation stage covered in chapters five and six. The DVD provides what the book cannot: dynamic media rich in colours and PowerPoint/Keynote DIY (do it yourself) techniques. The final task will be for you to visualise yourself presenting and answering questions from the audience. This part is essentially dynamic. Book chapters 7 to 10 as well as DVD tutorials build up your personal skills and confidence.

This book is great, as an *in vitro* tool. It is up to you to make it come alive *in vivo*, *in situ*, not in the test tube of a classroom, but in a real conference setting. The good news is that presentation skills can be learned. Among the hundreds of talented research scientists from A*STAR Research Institutes (Singapore's Agency for Science, Technology, And Research) who attended the presentation skills class based on this book, not one felt hopeless. And yet, the challenge they had to face was colossal. The presentation of each paper was only 7 minutes long. It took place in a large auditorium in front of scientists from different institutes. Engineers presented to Biologists, Chemists to Computer Scientists, Immunologists to Technologists. They had to be clear and interesting to people from extremely diverse background and convince the audience of the worth of their scientific contribution. As if that was not difficult enough, each had to go through a gruelling 8-minute Q&A after their presentation, where their authority, competence, knowledge, and nerves were tested to the limit. Under such harsh conditions, success was far from being guaranteed. But succeed, they did. It was possible in parts thanks to the many live rehearsals each had in front of a smaller audience made up of experts and non-experts. It was also because they spent close to a full day simplifying and illustrating the one point each of their slides made until it was understandable and legible to all. It was because they understood that, in scientific presentations, the presenter's role is to be a host who keeps a tight rein on the computer co-host. Finally, it was because they had been coached on how to face the audience, rephrase and answer questions from experts, non-experts, and troublemakers; questions that were sometimes relevant and sometimes not, sometimes hostile or incomprehensible and sometimes over-friendly.

It is their success that gives me the confidence that you can succeed too. To make learning fun, I pepper the pages here and there with the story of my scientist friend, Vladimir, a fictional character who excels at being a bad presenter. Actually, Vladimir is a mosaic, a montage, a collage, made up of hundreds of little bits and pieces of things presenters do that they shouldn't. He is the ultimate counter-example. If you sometimes recognise yourself in him, it is through pure coincidence, of course. And if your name happens to be Vladimir, don't feel bad, my Vladimir does not have the same last name. Have fun. It's time to turn the page to chapter 1. Happy reading!

CONTENTS

Preface	v
Part I: Content Selection	1
Chapter 1: Paper and Oral Presentation: The Difference	3
• The Spoken Word vs. The Written Word	4
• Collective Audience but Individual Expectations	7
• Captive Audience Trapped in Time and Space	14
• Imposed Pace and Rigid Slide Sequence	17
• You, Personality, Face, and Voice	19
Chapter 2: Content Filtering Criteria	24
• The Audience Expects the Presentation to be About Its Title	27
• All Contributors Expect to be Acknowledged	31
• Novelty, Applicability, and Time to Explain are the Main Content Filters	33
Part II: Audience Expectations	37
Chapter 3: General Audience Expectations	39
• No Disconnect	39
• No Strain	42
• No Boredom	49
• No Disregard	50

Chapter 4: Scientific Audience Expectations   53
- Digestible Scientific Content   53
- Believable Content and Credible Scientist   56
- Useful Scientific Content   62

Part III: The Slides   65

Chapter 5: Five Slide Types, Five Roles   67
- Title Slide — The Name Card   68
- Hook Slide — The Attention Grabber   73
- Map Slide — The Head's Up Option   79
- Story Slides — The Proven Claims   83
- Conclusion Slide — The Promised Items   91

Chapter 6: Slide Design   97
- Design for Slide Legibility   97
- Design for Audience Attention   124
- Design for Presenter Flexibility   135
- Design for Persuasion   144

Part IV: The Presenter   149

Chapter 7: The Master of Tools   151
- Screen, Pointers, Mikes, and Lectern   151
- Audio and Light Control   164
- Presentation Software (Keynote and PowerPoint)   164

Chapter 8: Scientist and Perfect Host   169
- The Attentive Host   171
- The Visible Host (and the Co-Host)   173
- The Hospitable Scientist   179

Chapter 9: The Grabbing Voice   182
- Speak with Confidence   182
- Speak for Intelligibility   187
- Speak for Attention   194
- Speak for Persuasion   196

Chapter 10: The Answerable Scientist **200**
- The Process of Answering Questions 204
- Three Troublesome Questioning Styles and How 217
  to Deal with Them
- Difficult and Dangerous Questions 222
- Typical Questions from Specific Groups 237
- Techniques for Fast Answer Support 239

Appendix **245**

Index **249**

Part I

Content Selection

Executive Summary

Content selection is guided by six criteria:

1) The expectations of people who directly or indirectly contributed to your work

2) The technical background required by an "imperfect" audience to follow your presentation (i.e. an audience in the same domain, but not expert in your field)

3) The expectations created by the keywords in the title of your talk

4) The novel or useful information that can be presented and understood quickly

5) Novel or useful information at a level of detail such that it convinces the audience to read your paper for more in-depth information

6) The task (either yours or that given by your management) to be accomplished via your presentation such as hiring, being hired, or securing funding

Listening to an author's oral presentation of his/her paper is inherently different from reading that author's paper (Fig. 1). The net effect of the difference is a transfer of responsibilities from the reader/audience (the information consumer) to the presenter/author (the information producer).

2 Content Selection

Role reversal in oral presentation (vs. paper)	From Reader/Audience	To Author/Presenter
Selection of content of interest in paper	----------->	
Control over time	----------->	
Control over environment (light, audio, heat)	----------->	
Control over information flow and sequence	----------->	

Fig. 1. The role of the presenter of a paper has expanded significantly compared to the role of the author of a paper. The audience is now less active, less "in control" than the reader. It has lost some of its prerogatives and initiatives.

1

PAPER AND ORAL PRESENTATION: THE DIFFERENCE

Compression exercise

Vladimir looked at his paper and uttered a deep sigh. Distilling twelve months of work into an eight-page paper was hard enough, but cutting his work down to fifteen slides was near impossible. He had so much significant data, so many formulas, figures and tables to present. He loved mathematics and was very good at it. He would dazzle the audience by moving from one complex formula to another with the greatest of ease. Therefore, he decided to keep them. He was also very good in modelling, and used complex but accurate numerical analysis methods neglected by other scientists. He would spend some time explaining the methodology, and the algorithms. That way, the results would make more sense, he thought. He scrolled down the pages to the results part of his paper and looked at his twelve figures. He needed to bring these into his presentation, but they would occupy at least three quarters of his fifteen slides. That was too many for his liking. He grabbed his chin and pinched it while thinking of a way to make them fit in less slides.

"What if I merged these two figures into one by adding a right vertical axis to the XY diagram to turn it into a XYY' diagram since both figures share the same X axis. Um, alternatively I could also shrink the visuals and put two on one slide." Vladimir was unsatisfied. He was convinced there was a better way. Suddenly, he found it! Instead of using figures, he would go back to the original data and put them inside several large tables that would fit in less slides.

"Yes! That will work!"

He felt very pleased with himself. The challenge of squeezing his large paper into a few slides no longer looked so formidable.

"Hey," he thought, "maybe there is also a way to present my formulas."

His mathematical and analytical skills could surely be applied to this greater challenge. He looked back to the methodology section of his paper, the one with the Navier Stokes equations written in the Eulerian Lagrangian formulation and thought... and thought... Somehow, presenting these formulas was not as easy as presenting data because there was a lot of text describing them. He used his finger to count the number of coefficients: a_{lf}, a_{tf}, c, C_t, f, h, h_0, h_{lf}, h_{tf}, k, t, t_0, Re, St, u, U, x_i, x_{fc}, x_{lf}, x_{tf}. There were 20 of them. He thought about shrinking down the size of the formulas but the subscripts would not be very readable. Hmm ... He pinched his chin a little harder.

"Maybe I could just skip some formulas and tell the audience to read about them in the Kim and Choi 2002 paper... No, that would not be good."

Suddenly, he smiled and snapped his fingers as his brain shouted a silent eureka.

"I will use one slide to write down the definition of all coefficients, and another slide to write the formulas."

Vladimir smiled. Another problem solved by the mighty brain of the great Doctor, he thought.

The Spoken Word vs. The Written Word

Poor Vladimir! Squeezing an eight-page scientific paper inside a 15-minute scientific talk is an Herculean task (Fig. 1.1). Just consider the physiological reasons:

At the rates given in Fig. 1.1, assuming one reads the whole paper, 45 minutes would be necessary to finish reading. This excludes the time necessary for the explanation of figures and tables. Let's say explaining

	Eight-page journal paper	15-minute scientific talk	Compression ratio
Number of printed words	~4500	500 max (50 words per slide*)	9/1
Speed required for comprehension	Reading: ~240 words/min**	Speaking: ~120 words/min***	2/1

Fig. 1.1. Compression factors to be considered to go from an eight-page written paper to a fifteen-minute presentation.
* Assumed here are one and a half minute per slide, a reasonable estimate given the time needed to explain the many visuals typically found in a scientific presentation. Beyond 60 words per slide, two unwanted side effects occur: 1) readability at a distance is greatly curtailed and 2) the audience starts reading instead of listening to the presenter.
** Observed over several thousand scientists with English as a second language (most probably auditory readers). English native speakers do better but, in a formal presentation, you have to give time to the non-English native speakers to read.
*** Speaking speed for conversational English is actually higher (165–180 words per minute). However, a formal presentation is not a conversation. The pace has to be slower to ease comprehension (particularly for a scientific presentation to an international audience, or from a presenter with a strong accent).

them takes another 15 minutes, bringing the total time to one hour. For a fifteen-minute oral presentation, a conservative compression ratio would be 4 to 1. This means that three words out of four have to be taken away when one goes from the written word to the spoken word. The compression ratio gets worse and exceeds 9 to 1 if one takes into account the restricted number of slides, and the restricted number of words one can write on one slide given the font size required to keep text readable on a projection screen. Vladimir is facing an impossible task. Cutting eight words out of nine is not cosmetic surgery. It is butchery. Cutting the fat (the details) is insufficient. One has to cut into the meat and into the bone and yet keep enough detail to convince others of a scientific contribution. Put as an information/time compression challenge, the problem seems intractable. Fortunately, the solution is not to be found in some optimisation scheme, but in "the inherent difference that exists between oral and written communication".

Intuitively, all agree there is an inherent difference but then again, there is also so much resemblance between the two. A paper is read. Shouldn't slides be read also? A paper has a standard structure that follows the scientific process from observation to hypothesis, methodology, results and discussion. Shouldn't a presentation follow the same structure? A paper contains sufficient detail to allow the reader to verify the claim independently. Shouldn't a presentation enable the audience to verify the claim? A paper presents facts in an objective, impersonal way to the reader. Shouldn't a presentation remain impersonal?

I discovered that the answer to these questions turned out to be a resounding NO. Why? Let us review these inherent differences.

The first inherent difference is in front of Vladimir: the audience. It is not miles away in another continent, reading his paper; it is right in front of him. Some people even attend his presentation for reasons other than to find out about his scientific contribution. Let's find out. Come with me inside the conference room and let us sit together on the last row next to the exit door where we have a good view of everyone entering and leaving.

Blinded

Vladimir is now behind the lectern. The notebook on the narrow ledge of the lectern in front of him displays the same slide as the large screen behind him, his title slide. The chairperson sitting at a low table to his right is reviewing the notes on Valdimir's profile before the formal introduction. Vladimir uses these few seconds of silence to glance at the audience. There are about a dozen people in the room. Three are sitting in the last rows, ready to make a fast exit.

Two are sitting next to one another.

Everybody else sits alone, mostly at the end of a row by the aisle, conference bag conveniently placed on the adjacent seat, preventing anyone to sit close by.

His manager sits in front of him in the second row.

A senior person sits in the middle of the room, directly facing centre screen. Vladimir is distracted by the entrance of a younger woman in a pale blue blouse that matches the darker blue of the conference badge holder clipped to the large buckled belt of her jeans. She picks a seat on the second row, not far from his manager who turns his head to look at her longer than he had planned to.

She is followed a few seconds later by a middle-aged man in suit and tie, the managerial type. As he walks down toward the front of the room, he notices the elderly person in the middle row and sits next to him.

They smile at each other and exchange a few discrete words. Had Vladimir been able to hear their conversation, he would have heard the following dialogue:

"Hi Bill, fancy seeing you here. What brings you to this talk?"

"Hi Chris. We sponsored Dr. Toldoff's research. That's why I'm here".

"Is he asking you for additional funds?" Chris asks.

"Don't they all…" He grins and asks "And what brings you here Chris?"

"I'm head hunting for my lab at Cornell. Dr. Toldoff's work is of interest to us. I'm checking to see whether Dr. Toldoff himself is also of interest to us. Would you transfer his NSF extension grant to our lab if we were to hire him?"

"It is worth a thought," Bill replies.

"No, it's worth millions!" Chris counters laughing.

The laughter attracts the attention of Vladimir who wonders who these two important-looking people are and why they came to his talk. The introductory words of the chairperson interrupt his thoughts.

Collective Audience but Individual Expectations

Some expectations are common to any audience, others are specific to a scientific audience, and others still are of a personal nature. For example, the sponsor of Vladimir's research is certainly expecting to see the name of his organisation mentioned on the slides (and that is only one of his many expectations). Let us look at the audience Vladimir faces and learn about their expectations. There are twelve people scattered in the room.

- Prof Christopher Mitchell-Untak, head of the speech lab at Carnegie Mellon University, reviewer of Vladimir's paper, hunting for new hires

Hi, call me Prof MU, my researchers do; I am the one who reviewed Dr. Toldoff's paper. Quite interesting research at the border between fluid dynamics and segmentation techniques. If we are right, these techniques could also be used in phoneme recognition, and that would be the major breakthrough we have been seeking for years to improve on HMM techniques. If Dr. Toldoff convinces me he has the skills needed for this work, then I'll make him an offer he can't refuse. What skills, you may ask? A clear thinker, good communicator, approachable person, knowledgeable expert are essential; stamina and dynamism would be good. I also hope his English is comprehensible.

- Postgrad student, with dissertation on parallel subject

Hi, never mind my name, it's as hard to read as to pronounce; I am writing my Ph.D thesis on the clustering behaviours of self-assembling fluid air-borne agents in the respiratory track. I wonder if my algorithms could apply to Vladimir's domain, or vice versa.

- Postgrad student, fascinated by Russian scientists, Vlad's potential wife

Hi, call me Tatania. I am a research assistant in Novosibirsk University in the department of advanced mathematics. My department head is a difficult person who diverts most of the department's grant money into his own pet projects, not ours. I was fortunate to have written a paper that was accepted for the poster session. I am single, and still looking for a husband, someone who speaks my native language. Vladimir is one of three such people attending this conference.

Collective Audience but Individual Expectations

- Dr. Gareth Linden, Vladimir's manager and co-author

Hi, and yes, with a name like this I cannot deny I am from that part of the United Kingdom called Wales. I co-authored Vladimir's paper. You see me here because Vladimir asked me to come just in case he is asked difficult questions during the Q&A. I did not want to be here because right at this moment in Room 15, the presenter is talking about a really interesting topic. Ah, well . . . I hope I did not come for nothing.

- Two researchers mentioned in the related work part of Vladimir's paper

Hi, I'm William Chatsworth. Dr. Toldoff mentions me in his paper. I'm interested to know if he has made any progress since he wrote his paper. I'm also wondering whether he will mention my work during his talk.

Hi, I am the guy to the right of William, the one with less hair. My work is also mentioned in the list of related works. I know Vladimir well. We co-authored a paper once, even though we belong to different institutes. There is one part of his paper I don't quite understand. He must have skipped a few essential details; I haven't had the time to talk to him about it yet. Maybe he will clarify it during his talk, and if he does not, I will ask him during the Q&A or later.

- Prof. Bill Smith, from the funding agency reviewing Vladimir's lab grant request

Hi, sorry for the suit and tie; they come with the job. My name is Dr. Bill Smith. I am from the NSF. I evaluate grant requests and Dr. Toldoff is Principal Investigator of one of the projects we sponsor. His lab has asked for more money. Since I am attending this conference, I thought I would find out from Vladimir how confident he feels about being able to succeed, and within which time frame. I will be looking at the confidence of Dr. Toldoff and how the audience values his work through the questions he is asked during the Q&A.

- Dr. Linda Woo, the potential boss of Vladimir's manager

Hi, My name is Linda Woo; Chinese father, American mother. I am a visiting fellow at the Chinese Academy of Science in Beijing to help a research group in non-linear non-statistical segmentation techniques. I have received a job offer from Vladimir's research Institute. If I accept the job offer, I will become the program director overseeing Dr. Toldoff's department, which is led by Gareth Linden. I have been told that both of them will be attending this conference, and I hope to evaluate them discretely.

- Researcher, interested only in the methodology used in the paper

Hello, my name is Tom Goh. I am a postgrad working at MIT as part of a university alliance program between MIT and my University. I do not really care about the contribution of Dr. Toldoff, but I am very interested in the method he uses to dynamically map his patterns based on the minimisation of a complex energy function.

Collective Audience but Individual Expectations

- Researcher, friend of Tom Goh, no particular interest in this talk, ready for a nap

Hi, I'm John, a friend of Tom. There is not one single talk that is of interest between now and three o'clock this afternoon. So, I'm hanging out with Tom. If the talk is interesting, I'll listen. If it isn't, I'll just nap.

- Researcher on industrial attachment, interested in the application domain

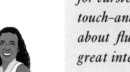

Hi, I'm Dottie, but people call me "Dot"; I am on an industrial attachment program with Motorola. My company is very interested in automatic segmentation of radicals for cursive handwriting recognition on the future Japanese touch-and-write phone, the jPhone. I don't know much about fluid dynamics, but handwriting recognition is of great interest to us.

- Postgrad, potential research assistant, bright but new to the field

Hi, I am a Biology postgraduate from Bangalore in India, with an additional degree in computer science; quite honestly, I am looking for a good place to work on interesting things. Dr. Toldoff's work is supported by an NSF grant. To me, that means interesting research with a budget. What could be better! Let's see if this Toldoff guy is someone I could work for.

How many of these people are so familiar with Vladimir's field of research that 1) they can skip his introduction, 2) they understand the value of his contribution, and 3) they are able to identify what is new and valuable in his research?

Half of the audience.

How many are only interested in certain aspects of Vladimir's paper, not the whole paper, and may require a certain amount of "hand-holding" to assess how the paper can be of use to them?

A third of the audience.

How many are equally or more interested in Vladimir than Vladimir's contribution?

Half of the audience.

How many are easily distracted during the talk?

A fourth of the audience.

How many people came to Vladimir's presentation because words in the title of his talk attracted their attention?

Half of the audience.

How many people also came to Vladimir's presentation because they hope to meet colleagues or friends from other research centres?

A third of the audience.

Who is likely to ask technical questions?

5/12 of the audience.

Who is likely to ask questions that Vladimir can't answer with authority?

Half of the audience.

What does this audience survey teach us?

1) You are presenting more than your contribution: you are presenting yourself to the scrutiny of others who will evaluate you, as a scientist, a co-worker, a research partner, an employee, an employer, a potential spouse, or as a financial risk. Therefore, the negative impact caused by a poor presentation far exceeds the failure of getting your point across to the audience.

2) Not everybody in the room is equally interested in your talk, but everybody entering the room has an expectation and hopes to see it fulfilled either during the talk, during the Q&A period, or afterwards.

3) Not all attendees to your presentation are experts. In our simulation, half of them are. The other half needs a solid introduction, and you are the one they rely on to highlight what is new because they do not have the knowledge necessary to identify it. The larger the conference, or the more general its theme, the more your audience is likely to need such guidance.

4) Contents should be tailored to meet audience expectations as much as possible. To the sponsor, sponsoring acknowledgment is important. To co-authors, co-authorship reflected in the title slide is a mark of respect and recognition. To those whose work you compare with your own, a fair comparison is expected.

5) Most attendees come because keywords in your title interest them. Second in rank are the people who are equally or more interested in you than in your scientific contribution.

6) The audience is likely to ask you questions that take you outside of your area of expertise and for which you do not have an authoritative answer.

7) The audience is likely to ask very technical questions, the answer to which may not be on your slides. You may be asked to clearly identify what is novel, and to clearly delineate the scope of your contribution so that its usefulness and applicability to other contexts can be assessed.

What we learned so far is that content which may not have been considered essential before (acknowledgments, authors, future works) may be essential. We also learned that your contribution is evaluated in two ways: the material on your slides (the science factor), but as importantly, your performance on stage (the human factor). Your personal dynamism and conviction will be as important as the strength of the facts presented. Finally, we learned that what brings people to a presentation are the presenter and title keywords.

Captive Audience Trapped in Time and Space

Besides the differences between reading and listening and between reader expectations and audience expectations, there is a third inherent difference between a written paper and a scientific talk. It has to do with who controls the way people spend their time.

Readers have total control over when to start or stop reading. The audience, however, has little control over when to start or stop listening. It is not easy to shut your ears. Furthermore, the audience has no control over the length of a presentation. It is decided by the conference organisers. Of course, like readers, people can always disengage and walk out, or work on their laptops, or take a nap, or SMS their friends, but that is not what they intended to do when they entered the room. Readers speed up or slow down reading according to the difficulty of the text. They can decide to read the same paragraph many times. The audience, however, has no say on how long a slide stays on the screen. Because time marches on, the contents of each slide have to be digested while still on the screen. There is no pause button, no screen freeze feature for those who need more time to examine a slide. As the saying goes, "time waits for no man". Therefore, each slide should be self-contained. It should not rely on any external support to be understood. Why? Because the presenter cannot count on the audience having efficient memory to recall the acronyms, numbers, curves and other scientific niceties found on previous slides.

Besides being unable to turn back the clock and return to past slides, the audience has no control over the presentation schedule. If the reader can decide to postpone reading after a heavy lunch, the presenter, on the other hand, has little or no say on when to present. The timetable is decided by the conference organisers. Is it morning or afternoon? Is it on the first day, on the last day, or somewhere in between? Is it right before or right after a break for refreshment, or is it sandwiched between two other presentations? The audience's readiness to absorb knowledge varies with these circumstances which are beyond the control of the presenter. Fortunately, the presenter knows a few weeks before the conference what the presentation time slot is. A talk right after lunch, or the last talk of the day cannot be delivered in the same manner as an early morning talk.

The presentation time is preset, but preset time does not mean that time cannot be reset. The conference timetable is not as strictly respected as one might think. Time easily drifts. Some are asked to shorten their presentation at the end of a morning or afternoon because a kind but undisciplined chairperson allowed an interesting presenter or keynote speaker to run over his or her allotted time, or simply because there were technical glitches, late presenter arrivals, or late refreshments. To catch up with lost time, rushing often occurs close to the end of the presentation where the most important slides are. As a result, a good part of the audience is left behind at the most crucial moment.

A presenter who ignores the clock often runs into the time reserved for Q&A. From the point of view of the presenter, this may not be so bad. After all, it shortens the Q&A, a time when scientists supposedly run their scalpel along the thread of the presented arguments, possibly revealing errors, uncertainties and contradictions. From the point of view of the captive audience, however, the Q&A is essential. The audience must recover the freedom to engage with the presenter and ask questions to better understand the work presented and its applicability to felt needs. Through a Q&A, the audience may even help the presenter by suggesting new applications, alternate interpretations, or by confirming findings. Therefore, time must be kept for the Q&A.

Listen to the "Keeping to time" podcast on the DVD

Consequently, manage time to present your slides within the allotted time. How? By rehearsing, and rehearsing some more because it is only through rehearsals that time-related problems are identified. Another way to manage time is by planning to complete the presentation ahead of schedule. Doing so gives you more flexibility to add a little information if you detect that you are leaving your audience behind. It is also less stressful: you are no longer racing against the clock.

Here are two techniques to help you cut down presentation time.

1) Use time savers. The greatest time saver is a well-designed visual. You have heard people say, "A picture is worth a thousand words"; this is even more so in science where visuals compare, describe complex relationships, reveal patterns, provide visual proof, and demonstrate far more convincingly than words. In addition, visuals are great for an international audience. They are universal and bypass language limitations.

2) Put less on each slide, and have fewer slides. Reduce content to centre it on the new, the interesting, and the useful part of your contribution. Since there is a direct relationship between presentation time and the amount of textual and visual information presented, less information is equivalent to more flexibility in time management. "Less is more" is not just a catchy slogan. In the context of presentations, it is good advice. You are well advised to apply this slogan to the amount of text on your slides. Less text

on a slide means less time for the audience to read the slide and more time to listen to you because people are not so good at doing both simultaneously. Robert Geroch reflects the voice of many who coach presenters when he writes, "the amount of information you emit is irrelevant; it's the amount you cause to be absorbed that counts."

In conclusion, because time is short and revisiting past slides to refresh someone's memory is not an option, each slide has to be self-contained, and complex arguments made over many slides have to be consolidated along the way to carry audience acceptance forward a step at a time. Because time is a serious constraint, time-saving visuals should be the rule, not the exception, and overall content should be cut down to leave time for the Q&A and to ensure that a moderately knowledgeable audience has time to understand. Endeavour to show how each slide is connected to the title of your talk. It aids memory and strengthens the cohesion between each part of the presentation and its central theme.

Imposed Pace and Rigid Slide Sequence

The fourth inherent difference between a written paper and a scientific talk has to do with the audience's loss of control over the flow and sequence of things. The audience cannot speed up the presentation, nor can it skip a slide to see what is coming after it.

Readers control their reading path in order to maximise their knowledge gains in the smallest amount of reading time possible. They only have to flip the pages of your paper to discover its structure. This direct access inside the paper is a time-saver, allowing partial reading of the paper and rapid identification of the sections of interest. Readers have total control over the way they wish to navigate inside your paper. They can jump directly to the results section, or decide to look at figures first, or skip the introduction, or first read the abstract and the conclusion before reading the rest of your paper, or come back to a difficult part after reading some related work online.

In a scientific talk, the captive audience sits through a carefully staged sequence of slides for a fixed period. It has lost its freedom of choice. It cannot skip the parts of lesser interest. It cannot return to past slides, and it cannot change the order of your slides. As far as the audience is concerned, this situation is as satisfactory as one-size-fits-all so long as it is size six.

Proper information sequencing to maintain audience interest and neutralise the impatience of your audience (particularly the knowledgeable audience) is particularly challenging. Your only option is to make each slide interesting. How is this done? Return to your research logbook (you keep one, don't you?) and take note of what really interested you at the time you were conducting your research, such as something that removed a research bottleneck, something unexpected, or something new that could be used in many areas. Novelty and need satisfaction are filters through which to sieve the full contents of your paper. Once the interesting contents are identified, organise them along a story plot that keeps the audience actively engaged.

Story plots help establish continuity. Continuity is essential. It allows you to prepare the audience to what appears on your next slide. With each transition, the story smoothly evolves from one slide to another. The audience usually accepts short stories (up to twenty minutes) without an outline because time passes by quickly when the plot is good. Longer stories (one-hour presentations for example) may require an outline to sustain attention over a longer period.

Stories are spun around the theme of the title. The story thread need not follow the traditional sequence found in a scientific paper: from introduction to background, from methodology to experiments, and from results to discussions. This is only one of the many possible story threads. Nowadays, rare are the movies that start with the title. Most movies start in action. You may want to consider starting "in action" to grip the audience with a compelling problem prior to introducing your title which contains a hint on how the problem will be solved. But if your results are particularly impressive or seemingly impossible to achieve, you may want to start with them right after the title to raise the attention of the audience to a high level. Similarly, if the potential application of your research is extremely interesting, you may want to start with it.

To conclude, because an audience, unlike readers, has no direct access to the insides of what is presented, contents are structured along a clear story that sustains interest. Each slide corresponds to one step of the story, naturally linked to the next slide/step. Each step contains at least one novel or useful element to moves the story forward. Novelty and potential usefulness to the audience become filters to sieve through the contents of the paper in order to identify the story elements. The story itself is centred on the title but does not necessarily start with the title.

You, Personality, Face, and Voice

The fifth and last inherent difference between a written paper and a scientific talk is YOU, your unique personality, voice and face.

Readers have access only to your name, an inked proxy, a pointer to the real you. Readers neither see nor hear you. Indeed, the passive voice makes you invisible so that facts appear objective. For example, in the expression "it was found", who is this anonymous "it": one person, a group of researchers, or several researchers from different labs? No one knows. What matters is what was found, not who found it.

In oral presentations, the audience gets to see you, hear you, smile at you, and engage or joust with you. Surely, YOU make a difference, or at least you should, shouldn't you? You have more channels of influence than just black text on white paper.

Oral presentations exploit the visual channel. In front of the audience stands the person who authored the paper. The authority of the author is going to be evaluated through his or her persona, look, confidence, smile, charisma, energy, and ability to answer technical questions satisfactorily. The authoritative presenter removes the doubts of the doubters. The non-authoritative presenter casts doubts on the quality and validity of the results presented. Authority is not just a matter of science. Visual clues such as posture, confidence, attire, make up, grooming, and even age, size, assertive gestures, and facial expressions, help the audience form an opinion on the authoritativeness of the presenter. Projecting an authoritative image of you to the audience is as relevant as projecting scientific proof on a screen. This authoritative image is reinforced during the Q&A session by the pertinence of the details contained in your answers. Clear, authoritative answers rescue the presenter of a poor formal presentation.

Besides the visual channel (the eye), oral presentations also exploit the auditory channel (the ear). On the positive side, despite average to mediocre looks, one may have a gorgeous warm voice, rich in tonalities and inflections, authoritative, evocative, probing, sometimes energetic and sometimes quiet.

On the negative side, one may be good looking but have an uncomfortably high-pitched nasal voice, or a voice so soft that nothing short of holding a microphone one centimetre away from the mouth keeps speech audible. One may have a thick Chinese or French accent only understandable by their fellow citizens. One may speak at Formula One racing speed, or at turtle speed. One may speak with a voice going up and down with the regularity of a pendulum or a flat horizontal voice keeping within a very narrow frequency range. One may have a popping voice that blasts air into the sensitive microphone with every explosive B or P, distorting speech and decreasing understanding. And one (hopefully someone else, not you) may even have a high-pitched, nasal, soft, accentuated, fast, and flat voice. Good use of our vocal chords is essential to succeed in oral presentations.

The voice — its loudness, pace, intonation, and accent — profoundly impacts how the audience understands and perceives your research. It helps you to emphasise, convince, charm, relax, attract, mark transition points, or wake-up an after-lunch audience. Voice, in synergy with the visual channel, increases memory recall and understanding. The absence of voice — silence — also improves understanding because it helps the audience to catch up with you, focus on you, read, or reflect on what you said. In some cases, silence also adds suspense.

There is an oft-disregarded factor that makes speech a great tool for scientific presentations. Intrigued? It is the oral factor. Something happens when you say things instead of writing about them. Your grammar changes: you use the active voice more often and become more personal, and you use more active verbs and verbs conjugated at the present tense. Your vocabulary changes: you use simpler verbs, less abstract vocabulary. Your sentence structure changes: you use shorter and even incomplete sentences. As a result, spoken language is by nature less complex than written language. It is also more repetitive. This facilitates understanding. The following example illustrates the many differences between written and spoken text.

WRITTEN: The sentences found in scientific journals tend to be long because the precision required for stating claims constrain the writer to use conditional provisos ("provided", "if", "when"), detailed qualifiers and modified nouns ("hydrophobic sol-gel hard coating"). (36 words — one sentence — 9 seconds to read)

SPOKEN: The sentences we read in scientific journals always seem to be *sooo* long, don't they? That's because we are scientists. We need to be precise. So we add conditions in front of our claims: "Provided this, provided that…" or "if this then that". Now, these additions make a sentence long. On top of that, we have to add modifiers and qualifiers in front of our words. It's not just the "coating" that peels off; it's the "hydrophobic sol-gel hard coating". These modifiers make a sentence that much longer, but of course, that much more precise. (95 words, i.e. more than two and a half times the number of words of the written text — 8 sentences — 46 seconds to speak)

What is different in the spoken piece? Did you spot the seven differences?

1) The conversational tone of the presenter is friendly. Use of contractions (*don't they, that's, it's*).

2) The presenter becomes part of the audience with the use of the pronoun *we*, and even interacts with it through rhetorical questions (*don't they?*)

3) The presenter seems to take a somewhat negative stance toward scientific sentences at the beginning of his speech, as if sentence length was a necessary evil (*sooo long*); but he balances that view at the end of the sentence by emphasising precision. Compared to the spoken version, the written version lacks contrast and is less persuasive.

4) Linking words, *but*, *now*, and *on top of that* are used as linking devices.

5) Spoken sentences are shorter.

6) The verbs are less sophisticated in the spoken piece (*seem to be* is less elaborate than *tend to be*).

7) Redundancies are many: *on top of that*, *add*, *also*, *that much*…

The reliance on two channels of communications, visual and auditory, removes the burden of proof from data alone to both data and presenter. A member of the audience still debating the validity of some projected data might be convinced by the assured tone and demeanour of a presenter confident that more data than that presented are available to strengthen the claims made. Of course, the audio channel may turn out to be a liability for presenters with weak oral communication skills, or for presenters whose assurance melts away in front of an audience. We will see later how to minimise or eliminate such liabilities.

22 Paper and Oral Presentation: The Difference

Does using the audio communication channel affect the amount of information presented? It certainly does. Speech is slower than reading. In addition, since everything on a slide has to be explained, there is a *de facto* total redundancy between the visual channel and the audio channel. For slides with visuals, the audio channel corresponds more or less to what is expressed in the caption of a self-contained visual (result and discussion). For a slide with text only, the audio channel is either a paraphrase, an expanded commentary on the text, or a straightforward repetition (a famous quote for example).

Apart from you, another factor directly influences the contents of your presentation: YOUR GOAL. I beseech you not to follow the crowd of scientists who go into a presentation with a mindset of untargeted information downloading. If you go into your presentation without a goal, there is no goal against which to measure your effectiveness. Set yourself a goal and let that goal affect your slide contents. What goal, you may ask? Let me suggest a few. You have been sent to the conference by your institute to present your paper but also possibly…

Listen to the "What are the benefits (of presenting)?" podcast on the DVD

1) To hire scientists. Where do you think your best prospects will be? They will be right in front of you during your presentation. Have your slides support this task too.

2) To make a career move (personality counts as much as expertise)

3) To find a sponsor for your research

4) To find an industrial partner

5) To find a research partner

6) To find a spouse (although I doubt your institute sent you to the conference for that purpose)

7) To get feedback and advice on how to improve your work.

In this section, we learned that presenters use their communication skills and personality as tools to influence how the audience perceives them and their work. They use the oral channel for emphasis and to reinforce the visual channel, thus making their work easier to understand, more memorable, more convincing, and more interesting. Presenters further sieve their contents through an additional filter: that of their own objectives.

2

CONTENT FILTERING CRITERIA

Dr. Sorpong's masterful preparation

Dr. Sorpong sat in front of his Mac laptop and launched Keynote, Apple's PowerPoint equivalent. It was still early morning. He preferred to do his hard thinking at the start of the day, after a hot shower and light breakfast. As Keynote opened, he did a quick calculation. For a fifteen-minute talk, assuming 90 seconds per slide, his presentation should have at most ten slides: eight story slides, plus the title slide and the conclusion slide. He did not bother using the research institute's compulsory presentation template at this stage. He had designed his own template for the task of preparing any presentation. The title slide was the regular "Title & Subtitle" master slide with a white background. The nine other slides (he called them his "story slides") had a top and bottom bar. The top bar, about 5 cm high, could contain a three-line sentence in font 36 — the story sentence. The text was white, and it cast a light shadow on a dark blue background. The bottom bar was similar to the top but the text cast no shadow and the bar could contain only one line — the title of his presentation. The title would be repeated on all slides; only the story sentence would change from one slide to the next.

Storyline

Evaluating Video Quality Metrics for Low Bit Rate Videos

He left the storyline bar unchanged but typed the title of his talk inside the bottom bar: "Evaluating Video Quality Metrics for Low Bit Rate Video". Going into the Navigator View, he clicked on that slide, held the option key and dragged nine copies of it into the slides window. He was now ready. Each slide had the title of his talk in the bottom bar, clearly visible. He would constantly look at it when creating his storyline to make sure everything the slide contained related closely to his title.

He went back to the title slide, and in the subtitle box, instead of typing his name, he started to write down the sort of people that might come to his talk, simply based on the keywords contained in his title. After all, the title words were the very words that would motivate the people to come to his presentation and create the expectations of what would be covered during the talk.

Evaluating Video Quality Metrics on Low Bit Rate Videos

- Experts in video quality metrics; low bit rate and high bit rate
- Experts in video compression Hardware/Software
- Experts in quality metrics, for example human interface guys working on subjective metrics
- Experts in telecommunications advocating variable rates of compression based on a mix of available bandwidth and quality metrics

After thinking about the audience, he realised that some people may lack video compression background, and even quality metrics background. However, he was sure that all had seen low bit rate videos and were familiar with the artefacts of highly compressed videos. With this in mind, he picked up his paper, and a green fluorescent highlighter, then sat back in his chair and started reading. From time to time, he interrupted his reading to highlight something novel or potentially useful to his audience.

> An hour and one cup of coffee later, he got up, stretched his body, and went to the coffee machine to refill his cup. He had read the whole paper and had highlighted enough content to start writing the headlines of the story described by the title of his paper. Each sentence contained in the top bar storyline introduced something new and useful. Each sentence was logically connected to the others through the story thread and summarised on the conclusion slide.
>
> For each slide, he would be asking himself the same questions:
>
> "How can I make this slide clear and interesting to the audience (aside from its scientific content)?"
>
> "What is the minimum amount of data I need to display in order to convince?"

The Audience Expects the Presentation to be About Its Title

The scientists in the audience have chosen to be in front of the presenter because certain keywords in the title of the talk attracted their attention. This very fact is of utmost importance. Because of it, and as incredible as it sounds, the scientific needs of the audience are predictable. Each title keyword attracts a segment of your audience. Identify these keywords, and you identify these segments. It is that simple. Let us use a sample title to identify the audience segments. Once you understand, do conduct the same exercise on the title of your own presentation.

A — Divide your title into clusters of keywords which correspond to knowledge domains (these domains may overlap).
Before practicing on your title, try your hands on the following title:

"Hydrophobic property of sol-gel hard coatings*"

---

*The paper was presented by its author, Dr. Linda Wu, at the 2nd International Conference on Technological Advances of Thin Films and Surface Coatings (Thin Films 2004) in Singapore. She kindly suggested some of the questions.

Content Filtering Criteria

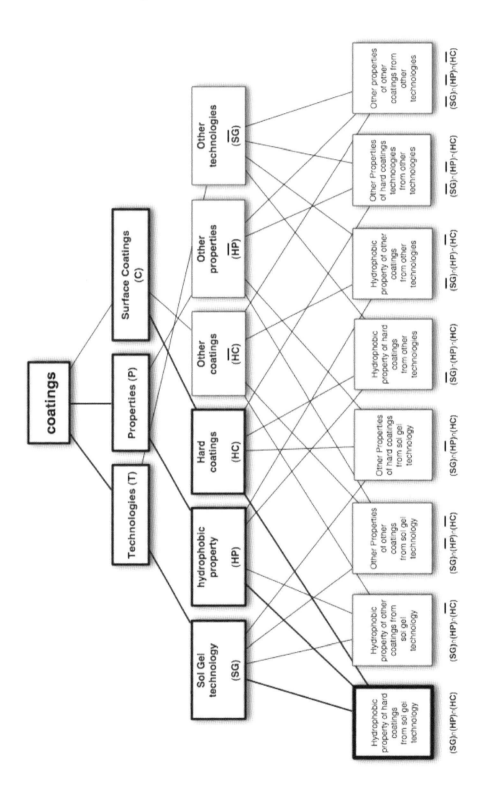

1) Sol-gel (SG): a subset of technologies for making coatings, T, which include other manufacturing processes, $\overline{SG}$

2) Hard coatings (HC): a subset of surface coatings, C, which include soft-coatings, etc. $\overline{HC}$

3) Hydrophobic property (HP): a subset of the properties of surface coatings, P, which include hardness, etc. $\overline{HP}$

Where:

$$T \text{ (Coating Technologies)} = (SG) \cup (\overline{SG})$$
$$C \text{ (Surface Coating Types)} = (HC) \cup (\overline{HC})$$
$$P \text{ (Surface Coating Properties)} = (HP) \cup (\overline{HP}).$$

The knowledge field presented in this paper is established by the three keywords in the title: (SG) ∩ (HC) ∩ (HP). Only the experts in the conference room have such knowledge. Naturally, the presenter would expect that the audience would be knowledgeable in SG, HC, or HP. But it should come as no surprise to also find among the audience people with knowledge profiles such as $\overline{SG}$ or $\overline{HC}$ or $\overline{HP}$. And it should also be expected that, given the name of the conference (2nd International Conference on Technological Advances of Thin Films and Surface Coatings), all participants are somewhat familiar with the main knowledge sets: coatings (C), coating technologies (T), or coating properties (P).

B — Once the audience knowledge profiles have been identified, identify the type of questions the audience may have as it enters the room where your presentation takes place.

Here are four possible audience knowledge profiles and their need:

$\overline{HP}$: What in the sol-gel process influences the hydrophobicity of hard coating? How is hydrophobicity related to other properties of the coating such as surface structure, transparency, or mechanical properties? How is hydrophobicity measured and quantified?

$\overline{HC}$: How are hard coatings made by sol-gel technology? How does hydrophobicity affect or confer the hardness of the coating? How hard is the coating made of sol-gel material?

$\overline{SG}$: What are the sol-gel precursor materials used? How are the sol-gel coatings solutions prepared? What are the coating processes used?

And (SG) ∩ (HC) ∩ (HP): What are the new aspects in hydrophobicity of sol-gel hard coatings?

C — Once the questions are anticipated, decide where and when they should be answered: on the slides, in the verbal commentary, or during the Q&A.

In our example, the contribution of the paper is mostly reflected by the keyword *hydrophobicity*. It is upfront in the title (as it should be). *Sol-gel* and *hard coatings* are placed after since they are part of the background. Therefore, information that pre-empts most of the background questions should be covered in the first part of the presentation.

D — In preparation for the Q&A, identify the type of questions that would be irrelevant.

For example, questions on hydrophobicity of soft coatings or questions related to hard coatings produced by other processes than sol-gel do not need to be answered because they fall outside of the scope of the presentation. You would also decline to answer questions on other properties of materials independent of hydrophobicity.

In this exercise, the title allowed you to identify six overlapping audience profiles:

1) People interested in sol-gel

2) People interested in hard coatings but only vaguely familiar with sol-gel

3) People interested in the hydrophobic properties of any coating material, whether hard or soft

4) People interested in the mechanical properties of sol-gel hard coatings anxious to know if the mechanical properties are compromised by the hydrophobic properties of the coating

5) People interested in the hydrophobic properties of sol-gel hard coatings

6) People interested in Linda Wu, the talented scientist

Because the title is such a powerful and selective audience magnet, do not just consider how well your title represents your contribution; consider also the type of audience it brings to your talk. For example, let us imagine that, instead of the title "Increasing hydrophobicity of sol-gel hard coatings",

Dr. Linda Wu had chosen a title such as *"Increasing hydrophobicity of sol-gel hard coatings by mimicking the lotus leaf morphology"*. She would have increased her audience size but at the same time would have had to be less technical because one of the new audience segments (the one interested in the morphology of plants) is less knowledgeable in sol-gel hard coatings. She would also have had to speak more on how morphology affects hydrophobicity.

Here are the additional audience categories her new title would have attracted:

7) People interested in hydrophobic coatings found in nature

8) People interested in the morphology of coatings

Therefore, when crafting your title, think AUDIENCE. Consider the various types of people it would attract, and adapt the contents of your slides and your verbal commentary to make sure each audience segment is able to follow your talk.

All Contributors Expect to be Acknowledged

Facing you in the conference room could be your collaborators (co-authors, research sponsors, or your employer). They have a stake in the outcome of your research. They contributed to it in some measure, but they are not the only ones. Indeed, your work is based on the work of many scientists and organisations without whom you would not have anything to present. As Pascal, the great philosopher and scientist wrote in his "Pensées"

> *Certain authors, speaking of their works, say: "My book," "My commentary," "My story," etc. They are just like middle-class people who have their own house and never miss an opportunity to mention it. It would be better for these authors to say: "Our book," "Our commentary," "Our story," etc., given that frequently, in these, more belong to other people than to them.*

Therefore, you want to make sure direct contributors or their organisations are acknowledged whether they are in the room or not. You can mention them orally or name them on the title slide or display the logo of their organisations. The advantages of mentioning them are legion. You display your social skills and your respect for others. You demonstrate that you value the work of others. You show that you are a team player (Vladimir has his future boss sitting in the room as well as a postgraduate interested to work for someone with a collaborative style and fairness of mind). If, besides featuring your contribution, your goal is also to hire new scientists for your lab, mentioning the research team is essential. I have seen acknowledgment slides displaying the research team with smiling faces all around. A sense of camaraderie and a pleasant research atmosphere emanated from the screen to fill the minds of many in the audience with thoughts such as "this is a fun group! I would love to be part of it" or "I would love to work there: dynamic team, exciting research!" Naturally, such slides require that the presenter's behaviour conveys the same message to the audience!

Some of the scientists attending your talk are competitors whose data or visuals may be displayed or compared on your slides. They want due credit given to their work. Your slides must contain the source of any visual presented that is not yours. Since this usually requires a full reference, i.e. much text, it is not always essential for the audience to be able to read it. Let the font size be small. People at the back of the room may not be able to read the source, but those in the front rows should be able to. It is appropriate to briefly mention the source of such documents orally to compensate for the lack of readability. Mentioning sources, particularly reputable ones, adds credibility to what you say.

 Check your title slide. Is your name readable in sans serif font like Arial or Verdana? Is your academic title, if any, mentioned?

The name is usually centred under the title of the presentation, in a smaller font size, and of the sans serif type. "*Sans*" is a French word that means "Without". If you don't know what a serif font is, look at any "Times" font. Notice the little feet at the bottom of most letters; they are the serifs. Serifs help form a line to facilitate the reading of small text. On slides, however, nothing is ever small, especially not your title or name. Therefore, use a sans serif font.

If there are co-authors, did you mention them?
Squash the fleeting thought of ignoring them. Unless you are the sole contributor to the science behind the paper, their names need to be mentioned.

If you mentioned the co-authors, how did you manage to make your name stand out without giving the impression that you are far better than the others?

Increasing the font size for your name, placing it on top while keeping the other names underneath looks boastful. To be discrete, you could use the same font size and style for all names, but underline your name or write it in a colour that stands out slightly more than the colour used for the others. For example, you could use a black colour font for your name and a grey colour font for the other co-authors. That way, you can even write the names in alphabetical order. Less discrete but effective also, once all names have been shown, use a custom animation to bring attention to your name by changing its colour, or by making your (smiling) photo appear while your name moves from where it was to place itself beneath your photo.

Is your presentation helping you achieve your goal of hiring collaborators or showing your future manager that you are an excellent team player?
In that case, you may even have a separate acknowledgement slide identifying the many who participate in your work: sponsors, co-workers, outside companies, your institute, and more. Naturally, you need not spend much time explaining that slide, but it will serve your purpose well.

Is it possible to use the logos of your corporate partners, including that of your own employer (research institute, university, government sponsor, research centre, company)?
Most research centres have a policy when it comes to the use of their logo. Check with them to find out what the policy is. Using logos instead of names on title slides (or acknowledgement slides if separate from the title slides) is preferable because title slides require a large space around their elements to avoid clutter. Many employers require that all slides display the company logo. We will return to this delicate topic later… but rest assured that the title slide should definitely respect company standards.

Novelty, Applicability, and Time to Explain are the Main Content Filters

"Audience" is a general term covering people disparate in knowledge and expectations. When it comes to novelty, what is new to some is old to others. Yet, the reason why each participant comes to your talk is identical: to learn something new about your title, or about you. The only person who actually knows what is new on the slides is you, the author of the paper. Novelty is usually dispersed in different sections of the paper: methodology, results, data set, or interpretation. Therefore, when preparing the presentation, going through your paper and underlining what is new and title-related is a good way to start content selection.

Something may be new but not yet have an identified use. Novelty in itself is interesting, but its applicability either immediate or future is just as interesting. Applicability is often highlighted in the abstract and conclusion sections of a well-written paper. It may even be worth highlighting secondary novelties, for example novelties in methodology, and secondary impacts, such as applicability of a new method to a different domain, particularly if you have time to illustrate them and if they relate well to your title. Going through your paper when preparing your presentation and underlying the usefulness of certain findings or methods is a good way to select helpful contents.

Unfortunately, the new or useful may also be complex. Formulas and graphics stacking a great number of curves take much time to explain clearly. As a rule of thumb, if anything (text, table, figure) in your paper requires more than one minute to explain to a non-expert, you have two choices:

1) You could remove it from your slides and put it on your technical slides. These are placed after your conclusion slide, ready to be used in Q&A.

2) You could simplify it. Focus on what is essential to make your point, and trim the rest (or keep it on a separate slide used for Q&A only).

Be aware that a visual takes more time to explain than you think. The legends of the X- and Y-axis are often difficult to read (small or vertical). The main point of the graphic may not be immediately clear to a non-expert audience and attention needs to be directed to the items of interest because the non-expert does not know what they are. Only a live rehearsal in front of a non-expert audience will help you determine how much explanation time is necessary, and whether or not you need to rework some of your visuals to make sure their main point is understood in seconds, not minutes (Fig. 2.1).

 Prepare three highlighters: green, yellow and red. Highlight in green anything in your paper (sentences and visuals) novel or useful that directly relates to your title. Highlight in yellow anything in your paper (sentences and visuals) novel or useful that does not directly relate to your title. Finally, highlight in red what requires removal or rework.

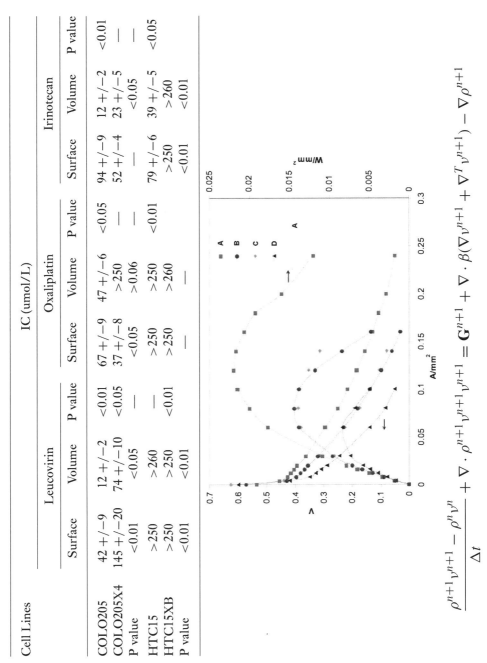

Fig. 2.1. Example of formula, table, and graphic that would take much time to explain.

Take the green highlighter, and whenever you encounter statements of novelty or impact directly related to your title, colour these.

By doing that, you may actually discover that some of the novelties were missed out or not emphasised in your paper. Maybe they can help you to make your case more convincingly. At this time, it is not necessary to be too selective.

Now take the yellow highlighter to surround any visual, figure or table that contains novel or useful material of secondary importance not directly related to the title of your talk.

Sometimes your paper contains many visuals that support your contribution. You do not have the time to present them all. It is therefore necessary to identify which one is essential to the claim made in your title. Secondary visuals are best left for supplementary Q&A slides.

Next go over the green-circled visuals again and use your red highlighter to strike out from them anything secondary, or to mark them as needing substantial rework prior to their use in your presentation. Three kinds of visuals require rework:

1) The visual (table, formula, graphic, diagram, or photo) that takes too much time to explain as it is, and thus requires simplification.

2) The visual that needs to be redone to make its point more convincingly. It may mean changing its form, for example a table into a bar chart, a dull photo into an interesting diagram with cutaway, or a static visual into an animated visual.

3) The visual that is not self-contained, i.e. one that relies on material defined in separate visuals or slides. Such visuals need to be modified to be self-contained.

Part II

Audience Expectations

Executive Summary

For the audience, the format of a lectern presentation is similar to that of a solo performance on a stage. The actor, in this case the scientist, relates to everyone in the audience, regardless of their knowledge, academic level, or nationality and accent. Frequent eye contact, the script of an interesting scientific story easily followed by all, and a deep respect for other people's work are three of the presenter's relational tools. The actor, now scientist, engages and guides the audience: makes it wonder, makes it think, makes it anticipate what comes next, and makes it value his or her scientific achievements.

Compared to the work required to read a written paper, people expect to accomplish more in less time by following an oral presentation. They know the presenter will focus on the essential and they are willing to forego a detailed justification as long as evidence and arguments are sufficiently convincing, and as long as the presenter is perceived as authoritative. People expect a talk easier to assimilate than a paper. The presenter fulfils that expectation by organising the talk in story form, by using animations to clarify, arrows, call-out boxes and colours to highlight, and by using a simpler language overall. People expect the talk to be easily followed. The presenter avoids losing the audience by making each slide self-contained and self-explanatory, and by speaking at a pace the audience can follow. Finally, people expect to be able to take advantage of their direct access to the speaker in order to network, to receive personal ad hoc scientific advice or opinion, or to understand better what, in the author's work, can be of immediate use to them. The presenter provides the means of contact — direct or indirect, immediate or differed.

3

GENERAL AUDIENCE EXPECTATIONS

When it comes to oral presentations, all audiences, scientific or not, have similar expectations. They expect to be able to follow the presenter and keep up with the story — no disconnect. They expect a mostly stress-free audiovisual experience — no strain. They expect to see someone interesting, someone who can make the complex simple and the uninteresting interesting — no boredom. And they expect to be recognised and to be spoken to — no disregard.

No Disconnect

Listen to the "Not so expert audience with distracting laptops" podcast on the DVD

Leaving people behind in your talk is easy. Effectively disconnect them from what you say. Simply zip past your title slide, knock them down with an overly complex slide, speak at the speed of light, glide over your slides without landing on them, and let chaos bring order to your presentation. The audience expresses these knock-outs with simple but deadly sentences a presenter never wants to hear: *It's too fast*, *It's not clear*, or *I didn't get it*. Let us review these five audience knock-outs, one by one, and learn how to avoid them.

1) **A rushed-through title slide** has a disastrous effect on the whole presentation. Before being able to follow your story, the audience has to fully understand your title. That takes time. At this early stage in your presentation, only brief introductory explanations are necessary such as the definition of a keyword you assume unfamiliar to the audience. Apart from explaining title keywords, another great way to spend time on the title slide is by rephrasing your title as a question and explaining the significance of this question.

For example, let's take the paper of Yu Han, Su Seong Lee, and Jackie Y. Ying, "Pressure-Driven Enzyme Entrapment in Siliceous Mesocellular Foam", published in *Chem. Mater.* 2006; 18: 643–649.

While on the title slide, the presenter might spend a minute and a half presenting the title and even have an animated visual to help explain it. Here is a possible introduction to the talk.

As you know, today, enzymes are used as catalysts to synthesise organic and pharmaceutical products. However, given their nature, they need a carrier, and not just any carrier, but a hydrophobic porous carrier to entrap them and keep them in suspension. The carrier will in part be responsible for the performance of these enzyme catalysts. If the enzymes easily wash out of its surface (it's called leaching), reusability is decreased. If the enzymes are not protected from surface heat, their structure gets distorted, and their catalytic effectiveness is reduced. However, if the enzymes are effectively restrained inside a pore of the correct size, leaching decreases and thermal stability increases because the enzymes are now locked into place and protected from surface heat. So our two questions are as follow: 1) Can we engineer a hydrophobic porous material whose pore size can be controlled to match a specific enzyme conformation, and 2) Can we design an enzyme loading process able to restrain the enzyme inside the pores to keep leaching to a minimum and thermal stability to a maximum. Today, I intend to demonstrate that siliceous mesocellular foam is such a material and that pressure-driven enzyme entrapment is such a process.

Once presented in this manner, the content of the rest of the slides is straightforward: the presentation has two parts, each one answers one question. The title is the structuring element of the whole presentation.

2) **Slide complexity** comes next in the list of presentation errors that leave the audience behind. The number of curves in a diagram, the number of diagrams, the number of bullets on a slide, and the number of words under

each bullet, all contribute to complexity. Numbers do not add up to produce simplicity.

Complex is an adjective. Adjectives are not objective; they are subjective. What is complex to the non-expert is simple to the expert with extensive prior knowledge. People in the audience unfamiliar with your narrow field of research consider formulas and overly detailed tables complex. It is best to avoid them, and by *them*, I do not mean the non-experts! They cannot be simplified, but your equations can. If the experts are hungry for more details, store these in supplementary slides only to be used to answer questions during the Q&A. Keywords familiar to you may be unfamiliar to others, therefore take the time to explain them in a just-in-time fashion on the slide on which they are introduced.

3) **Fast-talking** leaves behind the non-native English speakers. Sometimes, presenters are so nervous that their pace of speaking is faster than normal. Sometimes, they naturally speak fast. Combine the nervous talker with the fast talker, and you get the amazingly fast talker. To reduce fast-talking, calm down. Breathe deeply but slowly before you begin your presentation. Imagine yourself speaking to people outside of your field or giving a political speech. Listen to politicians — senators or heads of states — they never speak fast.

4) **Fast-pacing** is different from fast-talking. Here the presenter speaks at a normal speed, but goes through the material on the slide at a fast pace, not giving the audience time to digest. Overly ambitious presenters have more material on a slide than they have time to explain. They notice this during their rehearsal, but instead of reducing the number of slides or the amount of material on each slide, they just increase the pace, skimming over the material displayed, imagining that the audience is able to follow. It is pointless to leave text-heavy slides for the audience to read and for the presenter to ignore. With text-heavy slides, people stop paying attention to the presenter and read by themselves to gain knowledge.

Therefore, give yourself this rule: do not put on a slide what you do not intend to explain. Do not be caught saying, *You don't need to read everything on this slide, just remember that ...*

Fast-pacing often occurs two thirds of the way into a presentation when presenters realise they are running out of time. As a result, the key slides (the ones that highlight the contribution and are usually at the end of a presentation) are expedited, glossed over, or worse, read out without explanation.

Combine a fast talker with a fast pacer and disaster strikes. The audience is left behind in the cloud of dust raised by the galloping words. Avoid being in a position where you are forced to say, *given the time I have left, I will skip…* or *I wished I had time to describe this graphic in greater detail, but in the interest of keeping time…* Skipping story slides or gliding over them creates a gap between them and a conclusion slide that ends up claiming things unproven (as far as the audience is concerned) because the proofs were on the glossed over or skipped slides.

To avoid fast-pacing, talk to an audience during your rehearsal, even an audience of 1. People speak to themselves faster when they rehearse silently. If you do not have an audience, rehearse while speaking aloud. If you time your presentation through a silent rehearsal, you are guaranteed to exceed the time on the day of your real presentation. And when that day comes, talk to someone in the audience (preferably a non-expert) instead of talking to the screen or over people's heads; your pace will automatically slow down.

5) **Slide disorganisation** leaves the audience behind because people fail to piece the story together. A disconnected story is one where the audience is unable to see the logic of progression in your slides. It may seem logical to you, but since some in the audience do not have your depth of knowledge, they are not able to make the connections. The best way to detect such problems is to explain to yourself the logic of transition between one slide and the next. If you cannot, you will probably catch yourself saying *and next*, or *and here on this slide* during your presentation — a sure sign that you did not prepare your transitions. Make your transitions explicit for the sake of the non-expert audience. Imagine a question, the answer to which is the title of the next slide, and ask that question to capture people's attention.

No Strain

Ear and eye strain are the most common gripes of audiences worldwide. Said in a more polished way, audibility and legibility are important. The reader does not care about these because moving away from the source of noise fixes the audibility problem, and moving to a brighter spot or wearing glasses fixes the legibility problem. Beware of noisy conference rooms with poor acoustics (loud ventilation for example), and with poor lighting conditions or sun glare.

Beware of microphone pick-up patterns and of small fonts used to pack more data in slides already bulging with content. All create a whole range of audibility and legibility problems that spoil the enjoyment of the audience and are powerful turnoffs.

Memory-related strain comes next. The saying "into one ear and out the other" aptly illustrates the fleeting transitory nature of spoken words. People in the audience remember very little of what you have said during your talk. A famous communication study of UCLA Emeritus Professor, Dr. Albert Mehrabian (much quoted outside of context) confirms that the contents of your slides matter far less than you imagine. Have you considered how much of a scientific talk you can recall a few hours after the presenter left the lectern? Your answer is probably "Very little". You may, however, have kept a favourable impression of the presenter, and decided to read his or her paper, once back in your research lab. From the point of view of the presenter, this is a great achievement. Since the precise, lavish scientific facts on the slides are soon forgotten, do not vainly expect these facts to enter people's long-term memory. Focus on helping your audience understand your presentation by keeping your requests on people's short-term memory as small as possible.

Listen to the "What does the audience remember?" podcast on the DVD

In the book "*Scientific Writing: A Reader and Writer's Guide*" published by World Scientific, I dedicate an entire chapter to memory-related problems that impede reading. The same problems reappear in the context of a scientific talk to an even greater degree.

Here are the main sources of memory-related problems with which audiences struggle:

1) The acronym and abbreviation, particularly the ad hoc sort created by the author. It may have been explained on one slide, but as soon as the slide

disappears from sight it rapidly fades out of memory. The title slide often contains such acronyms. Avoid acronyms, unless familiar to all. If you have the slightest doubt, rewrite the definition of the acronym on the slide it appears or at least redefine it orally. The same applies to the symbols and names of variables used in equations. For example, do not say φy, but say *the magnetic field in the vertical direction*.

2) The pronouns, particularly *this* and *it*. Naturally, when you say *this shows*, you know what *this* means, but does the audience know? If *this* has a name, then say that name. For example, do not say *this shows*, but say *this decrease in speed shows*.

3) A reference to a symbol located on a previous slide is sure to strain memory. For example, let us say that in a formula on slide 6, you define φ as the magnetic field; slide 7 covers some other material, but on slide 8 you present a table (or a diagram) where φ reappears. Chances are that the audience has forgotten its meaning. Therefore, redefine it orally. Another unnecessary memory strain occurs when the presenter shows a figure and compares it to a figure already presented on a previous slide. Of course, the presenter remembers precisely ... but the audience rarely does. As you present, catch yourself referring to something presented on previous slides: *as seen previously in the XYZ diagram*", or "*this diagram, contrary to the one previously shown*, or *as seen before* ... Whenever you utter these or similar words, you can be sure to leave the following people behind: those with a bad memory, those with a good memory but who were distracted while you presented what you ask them to remember, those who came late to your talk and missed the first slides, and those who did not understand the previous slide you are referring to. Visuals must be compared on the same slide, side by side. Often, putting two visuals on one slide reduces readability below what the audience can accept (one may consider this a limitation of PowerPoint, as does Edward Tufte, a world expert in data visualisation).

 http://www.edwardtufte.com/tufte/index

Listen to the podcast "Powerpoint and Shakespeare"

Still, everything needed to make a point on a slide must be on the same slide. It is a severe constraint, but you do not have the choice. Here are a few techniques to help you solve such problems (note that jumping back and forth between slides is not an option).

a) Simplify the figures so that only the essential curves helping you make your point remain, and see if both figures can now be placed side by side.

b) Bring back, as an additional build (effect) on the same visual, the curve from the previous visual just long enough for you to make your comparison.

c) If both visuals share the same variable on the X-axis but have two different variables for the Y-axis, bring the two curves together on a XYY' diagram with a left Y- and a right Y'-axis (each with their own scale). This method works well as long as both curves are each identified by colour and by name, so that the audience knows which curve goes with which Y-axis.

d) Refresh the memory by displaying a scaled-down simplified version of the previous visual on your comparative slide. The audience will remember and compare more easily.

4) Lists of items are notoriously difficult to remember. Avoid unsorted, disorganised bulleted lists of items. Organisation increases memorisation because it reduces the number of items to a number of classes. In his interesting book, "*Clear and to the Point*" published by Oxford University Press, Stephen Kossyln presents eight principles one should follow to prepare good slides. The three obstacles presented under this heading fit neatly under "the principle of capacity limitations". But the list-of-items obstacle also illustrates his "principle of perceptual organisation", which

argues that things that are grouped into perceived categories are easier to remember than things that are randomly listed.

Maui Hotel

Vladimir had wished for a presentation skills class in Hawaii, hoping to enjoy a luau by a black volcanic sand beach watching wahines sway their grass skirts in rhythm with the undulating music of Hawaiian guitars. His wish had partly been considered by his manager. He was now in the foyer outside the ballroom of the Maui Hotel, in Detroit, enjoying a tasteless sandwich while watching penguin-like waiters go back and forth with trays of cups and saucers to the sound of bland elevator music.

When the class resumed, the instructor, a reformed four-star Michelin chef who had been seen on French Cable TV and claimed to have coached French President Sarkozy himself, announced that it was time for the "*lisez, tournez, parlez*" (read, rotate, and speak) exercise.

"You will one by one face the screen and read each gourmet dish on the slide that will shortly appear on the screen behind you. You are not to speak while looking at the screen", warned the trainer. "After reading, turn to face us, and without looking back, say what you have just read. This exercise will do three things for you. One, it will make your mouth water as you imagine these gourmet dishes. Terrified presenters have a dry mouth. Two, it will improve your memory skills, and force you to keep eye-contact with your audience while speaking because it is not proper to talk to people while turning your back on them."

Vladimir waited for the third point but it never came. Instead, the promised slide appeared and everyone in the room started laughing. The bullets were all impossibly-difficult-to-remember dishes that would never appear on an American restaurant menu unless it was run by a French chef.

Culinary Delights

- China Sea Calamari in fresh green Kampot pepper sauce
- French Goose Liver pâté on Gilroy garlic croutons
- Blanched almond pastry glazed with dark Helvetic chocolate
- Baked rib-eyed steak with horseradish and black sesame seeds

> Naturally, most failed to complete even one bulleted sentence without turning back to the screen. From time to time, someone would get two or even three bullets right and receive wild applause from the trainees.
>
> Vladimir the scientist tried to imagine French President Sarkozy complying with the instructions of the trainer. He failed. He also failed to see how this class would help him prepare his slides for the upcoming presentation of his research report in front of the institute's Scientific Advisory Board later that month — apart from learning that he had to avoid losing people through long bulleted lists of scientific gibberish as meaningless as these culinary delights.

Accent-related strain, it seems, does not fit under any of Stephen Kossyln's principles, yet it is a great obstacle that leaves most audiences behind. Everyone has an accent. And even within one country people have multiple accents, including the English and the Americans. French people easily understand English spoken with a strong French accent but they may be the only ones. It has taken me months to understand English spoken with a heavy Chinese accent. In a conference room, the audience, with only moments to get used to a strong accent, needs the help of the presenter.

Listen to the "Dealing with accent" podcast on the DVD

Here are four ways for the presenter to help the audience cope with a strong accent.

1) When is accent a problem? For things spoken, but not for things written. That is why people with an accent should help the audience by speaking less and showing more. Tables and figures speak for themselves as long as they are clear, readable and self-contained. A presenter's written English

is usually better than his or her spoken English. Mathematical, chemical, or other scientific notations speak a universal language understood by scientists all over the world.

2) People with an accent may not have the time to reduce their accent enough before they present. Nevertheless, they have time to rehearse their presentation in front of a native English speaker not familiar with their accent to identify the mispronunciations in the important keywords used on the slides. Failing that, they could record a native English speaker reading their paper aloud and then speak alongside the played-back recording to identify their own pronunciation problems.

3) Another way to reduce accent-related problems is to write the whole speech and practice reading it with a native speaker or a good computer-generated voice (Apple's Alex computer voice is outstanding — and comes free with every Macintosh). On the presentation day, the speech is read. It is not ideal, but it is better than uttering incomprehensible gibberish.

4) The presenter could attempt to familiarise the audience with his or her accent, for example, by slowly reading the title of each slide as it appears on the screen. This gives the audience time to match what is said/read with what is heard/read, and to decode the accent "on the fly". If you are online, visit the website

 http://www.reference.com/browse/wiki/Non-native_pronunciations_of_English

It is quite instructive on the matter even if you are not a linguist.

5) You could study English phonetics. It is easy. There are only 24 sounds for vowels and 34 sounds for consonants and semi-consonants to learn, 58 in all. Once you know them, any bilingual dictionary with the phonetic spelling of words will enable you to pronounce anything in the dictionary! For example, do you know how to pronounce the adjective "Aberrant"? It is pronounced æ'berənt. If this word looks unfamiliar, it is because it is written using the symbols (shown Fig 3.1) of the International Phonetic Association (IPA).

æ	as in b<u>a</u>g
b	as in <u>b</u>oy
e	as in n<u>e</u>t
r	as in <u>r</u>at
ə	as in tend<u>e</u>r
n	as in <u>n</u>ot
t	as in <u>t</u>ap

Fig. 3.1. Phonetic symbols for the word *abberant*.

Each symbol corresponds to one unique sound. Once you can voice these 58 sounds, you can speak any English word from its phonetic pronunciation.

Did you notice the apostrophe " ' " before the letter "b" in æ'berənt? It tells you to put the accent on the syllable that follows it, in this case "b". Therefore, you would say aBErrant. Handy, isn't it! You just need to buy a dictionary that has a phonetic transcription for each word, learn 58 sounds, and you are already half English!

No Boredom

The effect of boredom is worse than a temporary disconnect. It permanently switches off attention. What bores people and gives them the feeling they are wasting time is a perceived lack of progress. For example, the slide on the screen remains unchanged for five minutes while the presenter deviates from the slide topic or covers a sub-point in details. Another way to bore is to say things that are commonplace, or repeat yourself multiple times. Yet another way is to bring down eight bulleted lines of text, one at a time.

Do you know when an audience is bored? An audience betrays its boredom by raising the noise level in the room, or by losing eye contact with the presenter. If you attend a concert with abundant noise from blowing noses, clearing throats or shuffling chairs, you know the audience is bored. Lack of noise does not mean the audience is sleeping. On the contrary, the audience is in rapture, suspended to your lips because you are doing such a great job presenting. Listen for silence and look for stillness and eye contact to assess your audience's interest. After a humorous statement, look for grins or smiles.

Boredom, like curiosity, is contagious. If the presenter is bored giving the same talk repeatedly, the audience will sense it and also be bored. Therefore, treat each presentation as unique, as an opportunity to enrich your audience, at least in knowledge. Your latest presentation will then be as fresh as the first one.

Glitches, often a sign of lack of preparation, also waste people's time. They interrupt the presentation while the audience waits. Such glitches include the following appalling examples: clicking continuously on a button with a broken PowerPoint link to a video or an audio file hoping for a miracle to happen, apologising for the terrible text alignment on a slide because of a missing font in the lectern computer, browsing endlessly among slides to identify a slide of interest during a Q&A session, fiddling with the wireless microphone to turn it on, confusing the "next slide" button with the laser beam button, asking the chairperson which button to press to start the slideshow thus demonstrating that you are technologically-challenged, and to end this partial list, asking the audience if people can hear you while the microphone is obviously on and you are shouting into the mike.

Aside from these avoidable glitches, people also waste time when the presenter reads everything on the slides (not just the title). They could have read the paper in their hotel room or attended another talk.

No Disregard

Bonding with your audience removes its anonymity. It is best to give recognition to the audience, co-authors, chair, and sponsors early. During the Q&A, even if the question is hostile, say something nice when you recognise the person asking questions, as in *Professor Hamilton, it is nice to see you here today* or *Mr. Wang, I am familiar with the interesting work that comes out of your lab*. Do not belittle yourself as in *Professor Smith* (he is not a professor), *you should be the one standing here today*.

What makes an audience feel unwanted or ignored?

1) The following presenters excel at distancing themselves from the audience: the have-eyes-but-don't-see-you presenter, the presenter shielded behind crossed arms, the heliotropic light-seeking presenter facing the bright screen during the whole talk, the ghost presenter hiding in the shadow of the lectern. Such behaviour is often caused by stage

fright when presenters see themselves as victims in front of the firing squad instead of seeing themselves as hosts in front of guests (more on that later). One occasionally finds the speaker walking back and forth across the front of the room. Although this technique works for a rock artist on stage, it is inappropriate for a presenting scientist.

2) Besides lack of eye contact, presenters betray their unfriendly personalities by the tone of voice, and by words that are pretentious, obnoxious, sarcastic, uncaring, bombastic, arrogant, and narcissist. Some display their high ego upfront: *I presented this paper in front of the Australian parliament last week, and they expressed gratitude for my excellent work.* Some belittle the audience: *You probably are not interested in this highly technical research topic reserved to the few specialists who can actually understand this stuff, so I will make this presentation as short as I can,* or *You are supposed to know this material, so I will not cover it here.* As member of an audience, you have probably encountered such personalities. Some manage to hide their personality during the presentation, but come Q&A time, they reveal it. For example, someone's impatient nature will be revealed when asked a question that was clearly addressed in detail during the talk, and even explicitly shown on a slide. Such presenter shows impatience by answering *As I've shown before on slide...* thus indirectly scolding the questioner for not being attentive.

3) On the other side of the behavioural spectrum, one sometimes finds the overly friendly, overly chatty or overly comfortable scientists presenting with both hands (men) or thumbs (women) in pockets. They delight those who like this refreshing change in behaviour — a marked departure from the stiff corpse, bandaged mummy, or trembling leaf. But they also make uncomfortable those who expect a formal presentation, not a chat by the fireplace, and who find the lack of formality disrespectful.

To decide whether formality is expected, consider your audience and play to their expectations. Are these your work colleagues? Are you younger than most people in the audience? Are you a research veteran in front of postgraduates or a postgraduate in front of research veterans? Are you already recognised as a world expert despite your young age or someone whose citation count is less than three? Know your place, but regardless of your age, your knowledge, or your academic standing, show confidence. Confidence is vital to establish your credibility.

4) Besides your mannerisms, your crowded unreadable slides and lack of time control also send the audience the message that you disregard its needs.

5) The expectations of the audience are not just about the contents of your slides; they are also about you. Do not run away at the end of your talk. Some may have come because they want to meet you or because they know you. Others may have come because you are known as a good speaker with always something new and interesting to tell, and they want to make your acquaintance. Still others may have come with prepared questions for the Q&A session following your talk. If you forfeit your Q&A time because of an overly long presentation, they will be greatly disappointed.

4

SCIENTIFIC AUDIENCE EXPECTATIONS

What are the demands of a scientific audience? The scientific audience expects digestible scientific content, believable scientific content, and useful scientific content, in that order.

Digestible Scientific Content

In scientific presentations, knowledge is both a bridge and a barrier.
Knowledge of vocabulary: do you know what Metal Matrix Composite material means?
Knowledge of acronyms: do you know what UEVC stands for?
Knowledge of methods: do you know what the Hidden Markov Model is used for?
Knowledge of the Visualisation techniques commonly used in the field: can you read a Western blot? Is it the same as a Northern blot or a Southern blot? And knowledge of the domain where the field of specialty is found: to which domain does "electrokinetic" belong?

How does one ensure that contents are digestible? Ask mothers what they feed their babies. Once passed the stage of liquid food, the baby does not move on right away to solid food.

1) The mother adapts the texture of the food to the ability of the child. Carrying the metaphor into the presentation domain, it means that, early in the talk, the presenter must provide the non-experts in the audience with the knowledge necessary to fully understand the title; and later on, leverage on their increased knowledge to deliver content that is more technical.

2) The mother makes sure that a new spoonful of food is not given before the previous one has been swallowed. If she is not attentive, the baby will either choke or turn the head away to notify her. The presenter goes through each slide without rushing. Going through slides too fast is akin to force-feeding the audience. A force-fed audience ultimately switches off. Therefore, the presenter has to regulate the audience's intake of new information. An audience given novelty-rich slides requires a pause to chew and digest. A pause can be as simple as a short silence, or as elaborate as an oral summary acting as a transition to the next phase in the presentation. Like the mother, the presenter has to be attentive and monitor the audience for signs of restlessness, both visual and auditory.

3) In the metaphor of knowledge as food, bland text is hard and lengthy to chew, particularly in large chunks. Flavour-rich visuals are tastier. Although they also require chewing time and patient explanation, they are more nutritious and easily digestible when clear.

4) The mother no longer gives puree to a toddler able to digest solid food. Has the large number of highly specific technical keywords in your title already discouraged non-experts from attending your talk, thereby guaranteeing you an expert but small audience? Each conference attracts a particular type of audience. Is this a general conference? Is it a conference spanning cross-disciplinary domains like biomechanics where it is highly probable that non-experts will attend your talk? Is it a workshop for specialists only? Have you attempted to evaluate the level of expertise of your audience?

 How many conference topics does your conference offer? The greater the number of topics and the larger the attendance, the higher your chances are of having non-experts attend your talk. For example, consider IEEE Visualisation 2007 in Sacramento, California. It offered 19 conference topics to an audience of 730 scientists. (http://vis.computer.org/vis2007/)

Distributed and Collaborative Visualisation
Flow Visualisation
Information Visualisation
Isosurfaces and Surface Extraction
Large Data Visualisation
Multi-Resolution Techniques
Multimodal Visualisation

Novel Mathematics for Visualisation
Parallel Visualisation and Graphics Clusters
>>>Point-Based Visualisation
Security and Network Intrusion Visualisation
Software Visualisation
Terrain Visualisation
Time Critical Visualisation
Time-Varying Data
Uncertainty Visualisation
Unstructured Grids
Usability and Human Factors in Visualisation
Vector/Tensor Visualisation

As seen above, point-based visualisation is only one of the conference's 19 topics.

Now consider the 2007 IEEE symposium in Prague. Point-based graphics was its main topic. It had 100 participants.

Here is the list of its 11 sub-topics as found on
http://graphics.ethz.ch/ events/pbg/07/cfp.html

Data acquisition and surface reconstruction
Geometric modelling using point primitives
Sampling, approximation, and interpolation
Transmission and compression of point-sampled geometry
Rendering algorithms for point primitives
Geometry processing of point models
Topological properties of point clouds
Hardware architectures for point primitives
Animation and morphing of point-sampled geometry
Hybrid representations and algorithms
Use of point-based methods in real-world applications

Would you expect to see more non-experts attending a talk at the Sacramento conference?

Examine the list of topics for your conference carefully. It gives you an idea of the background of the participants. Put yourself in the shoes of participants specialising in a different topic than yours; would your presentation also be clear to them? What level of knowledge do you require from the attendees to your talk? How does that reduce the impact of your presentation (the higher the knowledge required, the lower the non-expert audience satisfaction)? Is your title written to reach the participants you want coming to your talk? What have you done to ensure that the participants you want coming to your talk are not left behind?

Believable Content and Credible Scientist

You need to be credible and your conclusions need to be believable. Both are necessary, and both are deeply intertwined: if one fails, the other fails too. The conclusions are based on your data, your assumptions, and the logic of your arguments.

Believable Content

Among the words that follow the pair of adjectives *believable scientific*, one usually finds the words *fact*, *evidence*, or *data*. To scientists, these words do not automatically inherit a high believability label. Data (yours or peer-reviewed data from other papers) lead the audience to believe that you have a *believable scientific . . . basis, theory, model, hypothesis* or *speculation*. Therefore, your first job is to convince the audience that these facts, evidence, or data are indeed *believable*. That label is hard earned. Stephen Senturia, a senior editor who frequently reviews scientific papers in electrical engineering*, points out what an author should do:

> *You should trace by example how you go from raw data to reduced data to extracted measured result, and mention such things as calibration [. . .], the number of samples, and the relation between the error bars on the graph and your data (is it full range, probable error of the mean? What?)*

In other words, to be believable, data need to be measured, reduced, verified and verifiable, sufficient to be statistically significant, and have their relevance explained. As a writer, you understand that the journal format provides more room for words than slides do to describe your data. As a presenter, you know that establishing believability in front of an audience takes time, and time is

---

*Senturia, S.D., "Guest editorial how to avoid the reviewer's axe: One editor's view", *Journal of Microelectromechanical Systems*, Vol. 12 issue 3: 229–232, June 2003, IEEE/ASME.

precisely what you are lacking. You even have to simplify some visuals to focus on the essential in order to shorten explanation time.

Stephen Senturia recommends writing "*the paper in order of decreasing believability*" and to use "*high reliability starting points*". This also applies to scientific presentations. If you claim something that is clearly biased or exaggerated in your first slides, the audience will immediately raise its guard and distrust you. Whatever you present next will be not be highly believable. As a proverb in the Ecclesiastes says, "*Dead flies make the perfumer's ointment give off an evil odour*[†]."

Besides being founded on reliable and relevant data, believability also rests on the soundness and relevance of the arguments, assumptions, and methods used to extract new knowledge out of the data. For example, a method may require data to follow a Gaussian distribution to have believable results. Is the distribution of data Gaussian? There will come a time during the presentation when results will require explanation. You will then propose a plausible interpretation. It will be all the more believable if you have managed to firmly secure the audience's belief, up to that point.

Complex multi-step arguments and methods may take several slides and the audience needs to suspend disbelief until the last of these slides. The great philosopher and scientist Pascal said that "*memory is necessary for all the operations of reason*[‡]", and it certainly is. To follow a complex argument/method, you must help the memory by first explaining the argument's logic and step sequence. Once the audience is convinced of its soundness and knows its steps, it can follow an argument and believe it more easily.

For example, imagine that you want to demonstrate that Gene B needs to bind at a specific Gene A binding site for a phenomenon to occur. This demonstration requires three steps.

1) Gene A without Gene B is unable to cause the phenomenon to occur.

2) A mutated version of Gene A without the specific binding site is unable to cause the phenomenon to occur when Gene B is present.

3) When Gene A binds with Gene B, the phenomenon occurs.

Instead of hoping that the audience is able to follow the thread of your argument as you progress along its logical path, it is safer to prime the memory

---

[†] Bible, Revised Standard Version, Ecclesiastes, 10:1.
[‡] Blaise Pascal, Pensées, Section 6 — thought 369. Editions Leo Brunschvicg 1897.

58   Scientific Audience Expectations

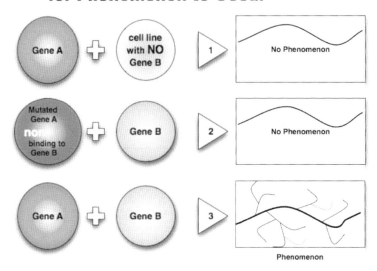

Fig. 4.1. Presentation of the argument structure prior to supplying the detailed evidence helps the audience follow a complex argument across many slides without difficulty.

Fig. 4.2. Drawing placed on the first slide following the argument structure slide to help the audience remember the step described on the slide.

and present the full argument graphically on one slide before presenting each of its steps on the subsequent slides. Here is a graphical example of the Gene A–Gene B argument (Fig. 4.1).

On each slide, a scaled down image of the step presented refreshes the memory of the audience (see Fig. 4.2).

To the reader of a paper, believability is the result of a critical analysis of the evidence, assumptions and methodology. The path followed to reach believability is not as linear as the author thinks. The reader may start with the results, but may also start with the references or with the introduction. Often, the reader (and reviewer) will move back and forth between results, discussion, and introductory statements, in essence walking through the paper multiple times. An audience has no such luxury because the presenter follows a unidirectional sequential path through the information. Therefore, believability has to be secured at the level of each single slide in the presentation. Conviction, particularly in science, is built on the accumulation of related concurring facts. These facts need to remain in memory as conviction strengthens. The audience finds it hard to keep all these facts in mind, particularly during a long presentation. Therefore, it is good to pause at times and remind the audience of the progress made toward your conclusion.

In summary, believability is established through what the audience perceives as quality data: representative of what is studied, statistically significant, suitable for the method used, and relevant to derive the stated conclusion. Over time, believability in the presented work solidifies and turns into conviction through the logical strength of the arguments/methodology. This gradual process requires support from memory and a systematic consolidation of the conviction slide after slide. Believability is destroyed at the slightest hint of bias or exaggeration in data, or perceived weakness in argument, assumption, or method. Data is necessary but not sufficient to secure believability, for the simple reason that time does not allow all data to be presented, thus forcing the presenter to select data based on their convincing power, and to enhance believability through independent means.

Credible Scientist

Here is another word that often follows the pair of adjectives "*believable scientific*": the word *professional*. When time is lacking, your professionalism (or your reputation as a talented researcher if you have been in the same field for some time) has to come into play to add believability to your data. Professionalism is intangible, qualitative. It has to pour out of you, your respectful yet authoritative posture, the clothes you wear, the confidence you demonstrate, and the clear slides. Data are believable, but even more so when presented by a professional whose reputation and whose organisation's reputation are at stake. The confidence you have is not just a matter of feeling good or looking good. Your confidence comes from your long experience in the field, your knowledge of methods and your knowledge of other people's results. It comes from months of scientific work, over which you carefully collected, cleaned and analysed your data, checked and refined your hypothesis, tackled and solved one problem after another. Confidence comes from knowing that not one person in the room, be it a Nobel prize winner or a faculty Dean, has conducted the research and has your depth of knowledge on your topic. Your confidence comes from the name and reputation of your organisation, and from the name and fame of your sponsors. Braided together like the strands of a rope, these factors strengthen your confidence, a confidence from which your authority and credibility emanate. As a writer, you may exude confidence and believe in your results and plausible explanations, your words may carry your conviction, but you cannot physically communicate your confidence to your readers. As a presenter, your words must convey and your behaviour must display your confidence to your audience.

Believability comes from perceived objectivity, honesty, and lack of bias. The French have a saying that describes suspicion born out of overexcellence: "*C'est trop beau pour être vrai*" (it's too good to be true). For the sake of intellectual honesty, the presenter should state, even in passing, the factors that limit the scope or applicability of the research. Honesty demands that deviations from the expected be recorded and presented even if they cannot be explained. When data blemishes have carefully been "cleaned-up" to remove whatever conflicting data remain in the path of stellar results and conclusions, one gets suspicious. Have data been manipulated so that an hypothesis can be verified? The Latin root of the word "fiction" is "*fingere*", which means to shape, and form. Reshaping data, in some cases, turns them into fiction. An audience uncomfortable with your choice of data may wonder if you have a hidden agenda or ulterior motives, and start questioning your intellectual honesty. If you get questions on your data during the Q&A, it may be that some people disbelieve your results or conclusions because they do not trust your data. To pre-empt such questions, establish the relevance and quality of your data during your talk.

Listen to the "Stating limitations" podcast on the DVD

Stephen Senturia recommends to *"make clear you are jumping off the believability cliff by making an assumption that is not probably correct."* This is not just wise advice; it is a way of displaying your intellectual honesty. Believability is based on the audience perceiving your intellectual honesty. It is also based on your fairness toward the work of other authors. Be fair toward others when you compare your work with theirs. These authors may well be part of your audience!

For writers, their reputation as experts grows as they write more published papers. Believability and credibility are attached to their names in the course of time. For the presenting scientist, reputation is gained immediately through an engaging presentation and a masterfully handled Q&A. There, in front of your peers, with no place to hide and no time to defer anything, you communicate your genuine scientific skills, communication skills, and social skills.

 How believable are you?

- Do you have a long track record of publications in your field? If so, does the audience know? Shouldn't it be revealed in your introductory statements with assurance but modesty?
- Do you need to compare your work with the work of others to highlight your contribution? If so, have you treated them fairly, have you attributed to them the visuals you borrowed from their papers, have you expressed gratefulness for their work in advancing the state-of-the-art in this field? Did you compare your data with theirs on comparable grounds?
- Do your visuals display error bars if known? Do you mention how you achieved quality data (calibrated equipment, standard cell lines, purity of solution, etc)? Can you explain the source of the errors, the kinks in your curves?
- Besides emphasising the strength and applicability of your method, have you also honestly revealed its limitations?

How memorable are your slides?

- Do you often feel like saying "as seen on a previous slide" during your presentation? Shouldn't you bring material from that slide forward again instead of relying on people's memory? Is there a way to avoid the need to

revisit old slides by restructuring the slides so that each slide is believable and convincing on its own, and the audience only needs to carry that conviction forward, slide after slide?
- Do you sometimes take a break in the relentless march forward through your slides in order to refresh people's memory, to tell them what has been achieved so far, and what are the next steps leading to your conclusions?

Useful Scientific Content

All attendees of a talk come with a goal: to learn something new and useful. These are adjectives, and therefore subjective when put in a collective context. What is new and useful to you may not be new and useful to someone else. The more widespread the interest and the more varied the knowledge of your audience, the more likely your presentation contains information that is new to some but known to others. Fortunately, the ones who already know are usually willing to listen patiently to familiar material as long as they are also given new material.

The challenge rarely comes from having to feed the experts new technical material. The real challenge is to make new concepts simple to understand by non-experts. This requires a clear mind not afraid of using metaphors or cross-domain illustrations, and the ability to identify early the topics a non-expert audience is likely to struggle with. During a recent cross-disciplinary presentation, a scientist was faced with the task of explaining protein motif-pairs to a group of computer scientists. She went on to explain that, like protein motifs, people are often seen in pairs, a father and a mother for example, because they live under the same roof; however, people are also frequently seen in pairs even when they do not live under the same roof, like a doctor and a nurse, because they share the same function. The metaphor worked. The talented scientist had been able to explain complicated things by relying on what the audience already knew. The audience was delighted to understand what seemed foreign at first sight.

Scientists come to listen to you because they hope you have something useful for them. Newness is pointless without usefulness. New information is useful when it creates, changes, refines, consolidates, or enlarges a viewpoint (Fig. 4.3). Maybe they also came with questions they needed answered to get clarity on your paper: questions on experimentation details, or questions on

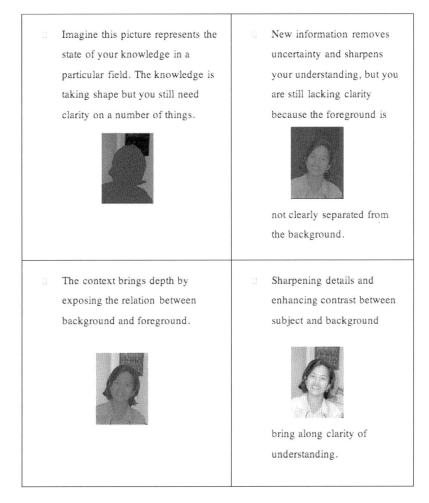

Fig. 4.3. To take a photo metaphor, useful information is equivalent to focusing on a subject, bringing in the background by increasing the depth of field, and working on the lighting to enhance contrast and brightness so as to bring the subject forward, without having to change or add to the background.

how far they can stretch the scope of your application to fit their needs. Maybe they came to find out why they failed to observe what you have observed (what they missed in their own experiments).

Useful scientific information is not limited to your slides. In a public presentation, a personal face-to-face communication channel is opened between the members of the audience and the presenter. This opening extends beyond the presentation room and beyond the time allotted to your presentation. Successful scientists use conference settings to engage in networking with potential employers, employees, sponsors, authors, peers, and collaborators, to establish research partnerships, suggest globe-spanning joint publications with other talented scientists sharing your research goal, or gather unpublished information such as research directions, and sensitive information such as intellectual property. Networking with you, may be an item on the to-do-list of some scientists attending your talk. It is your responsibility to facilitate this networking while being vigilant to keep confidential what is confidential.

Networking is prepared prior to the conference. It starts with the creation of your personal webpage, accessible by anyone anywhere. It continues with the printing of a set of colour copies of your slides available to the people who attend your talk. To each printed distributed handout, you have stapled your business card. It states your webpage and email address, as well as your institute and your abbreviated academic level. At the end of your talk, possibly right after you finish presenting your conclusion slide, you may have the opportunity to briefly present your ongoing or future research if you deem it likely to generate good networking opportunities.

 How useful is your presentation?

- For each slide, identify what is new for which audience segment, and then ask yourself the question: "so what?" What could those who learn about this for the first time do with it? Why should they care? Am I answering anyone's needs here or just intoxicating them with an overdose of meaningless new facts?
- Are all audience segments (from the expert to the non-expert) given useful information, in a balanced manner?
- Useful networking information is found outside of the presentation itself (your business card, your handouts, etc). How do you intend to expand the scientific impact of your presentation through networking?

Part III

The Slides

Executive Summary

The slides that make up a scientific presentation belong to three clusters: introduction, body, and conclusion. The introduction contains the title, hook, and map slides; the body contains the story slides; and the conclusion, unsurprisingly, contains the conclusion slide.

The title slide is the name card you hand to the audience, following the usual protocols that accompany such exchanges. The hook slide is your opening to grab attention, and secure interest toward you and your topic. The map slide, an optional slide for longer presentations, reveals the different parts of the story and manages time expectations. The story slides assemble a logical story woven around sentence-long claims visually substantiated on the very slides on which the claims are made. The conclusion slide ties in with both title and hook claims by reaffirming your scientific contribution and its foreseeable application in two or three points accompanied by their corresponding memory-enhancer visuals reduced to stamp-size.

Since technological and physiological limitations hinder vision, it is necessary to design legibility into your slides for text (adequate font type, font size, colour, spacing between text and visual objects) and visuals (contrast, resolution). The need for legibility is clear. The need for attention-getting, attention sustaining devices within each slide is probably less obvious to the new presenter. Physiological limitations of the eye, ear, and brain affect attention-demanding scientific oral presentations. Coordination between oral and visual communication is required to avoid communication channel conflicts. Fatigue raises the need to re-engage the temporarily disconnected brain using techniques such as questions, visual emphasis, expectation building, and eye-catching animations. Besides compensating for the inherent physiological limitations in any human audience, slide design also helps manage time by ensuring that certain slides may be skipped or inserted on the fly, for example through the use of hyperlinks.

5

FIVE SLIDE TYPES, FIVE ROLES

Click to add title

Vladimir stared at the blank PowerPoint slide, looking lost in front of the two familiar sentences on his computer screen: "Click to add title" and the smaller "Click to add subtitle". Next to him, under the desk lamp, laid his published paper, and somewhere in a folder on his hard disk were the PDF or jpeg files of the published figures ready to copy and paste into his slides. He had done the title slide. That was easy: he just had to fill in the template provided by his organisation. The title slide template had no area for acknowledging sponsors, so he did not mention NSF, his sponsor. He had prepared an outline slide. Although he wondered why an outline slide was necessary for a fifteen-minute presentation, he had banished the fleeting thought. He had seen so many presentations starting that way that he was convinced it was correct. He had quickly created four slides with the headings "Motivation", "Methodology", "Results", and "Conclusion", and one, the last one, featuring a large THANK YOU and Q&A. Being told to plan for one minute per slide, he counted on having fifteen slides after discounting the title and "thank you" slides, which he intended to display briefly. He looked at his paper and uttered a deep sigh — so deep that his wife Ruslana noticed.

"You need help?" she asked.

"I need a magic wand and a cup of strong coffee."

"I am your good fairy. So, what do you want to turn your cup of coffee into?"

"Discard the wand," he growled, "I'll settle for strong coffee and sleepless nights instead."

> "Hum, coffee that strong is hard to come by these days. How many sleepless nights do you want? I'm asking because I need to adjust the dosage."
>
> "Why in heaven did I marry a chemist!" Vladimir replied, laughing.

This chapter covers five types of slides: title, hook, map, story, and conclusion. Each of the five sections has the same two-part structure. The first part provides an example of the type of slide and explains its role in a presentation. The second part, written in question and answer form, acts as a design guide to help you create the best possible slide of this type.

Title Slide

- **Your title**
- **Your name**
- **Your company name and logo**
- **Acknowledgments**
- **A visual** if hook is on title slide

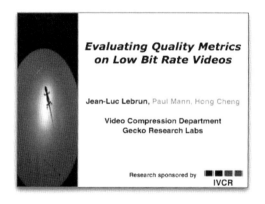

Title Slide — The Name Card

A — The title slide introduces the audience to contribution and contributor(s)

1) It helps the people in the audience recognise they are in the right room prior to the start of the presentation: the title of the talk is the same as the one in their printed programme.

2) It reminds the audience of the presentation topic and suggests or reveals the scientific contribution.

3) It reveals your identity, the name of your organisation and sponsor(s).

4) It indicates whether the research originates from people within the same organisation or from multiple researchers from collaborating labs.

5) It alerts the audience on the confidential nature of the information disclosed, except in conferences where the information becomes public knowledge, confidential warning or not!

B — The title slide gives the audience time to settle, relate to you, and assess your personality

6) In the first 30 seconds of the presentation, nothing more important could take place than to let the audience discover you, the presenter: how you look, how you sound, how you behave toward them (greetings, smile, attitude, humour …). There is not much to read on a title slide in any case; therefore, the audience will mostly look at you. Because the audience needs time to settle down and relate to you, do not rush past the title slide.

7) Some people require a little time to adjust prior to giving all their attention to any given task. Once they are tuned in, however, they focus fine. Such tuning requires time, just as it does when locating a station on a short wave radio. If the signal is strong, it is fast. If the signal is weak, it takes longer. Therefore, start strong, loud, and clear.

8) Your friendly personality and team player quality are revealed through the brief and appreciative presentation of your research collaborators.

C — The title slide enhances your credibility and the credibility of your results

9) Some labs have international name recognition in certain fields. Belonging to such labs is, to some people in the audience, a sign of research quality.

10) Collaborative projects funded by reputable sponsors encourage a presumption of independent validation of the research results by all collaborators to the paper.

D — The title slide promotes name recognition and branding

11) Most organisations encourage (and sometimes require) the use of "corporate" presentation templates: one for the title slide, and one for the normal slide. The consistent use of these slides by all their researchers creates and promotes branding. Logo recognition is of prime importance to any organisation.

E — The title slide may also serve as an attention-channelling hook slide

12) Keeping the hook on the same slide as the title slide has its advantages. It helps explain the title and gives the audience more time to focus on the topic of the presentation. If the hook requires much screen space, however, it is better to keep it separate from the title slide for legibility's sake.

Frequently Asked Questions

Should the title slide reflect the name or date of the event?

From the point of view of the audience, this information is not useful. The more information you put on the title slide, the more unpleasantly busy it looks, and the less readable it becomes. Unless you are forced to use the presentation template provided by the organisers of the event, do not succumb to the temptation.

From your point of view, it is helpful to mention the event date and name on your presentation, so that you avoid presenting the same things twice at similar events. However, for the sake of keeping your slide clear, add date and event to the title slide once the public presentation is over, not before.

From the point of view of the organisers of the event, it promotes event name recognition, and therefore helps them. Unless your aim is to also promote the event, you may want to pass on the opportunity of using the event slide template, while taking note of it because it may have been designed with the audience in mind (legibility, proper fonts, etc).

*Can I acknowledge sponsors on the title slide instead
of on the acknowledgment slide?*

At the start of your talk, the audience needs time to get used to you, and the time to acknowledge others can be used for this purpose. It is better to do the acknowledgements early, particularly if those acknowledged are in the audience. They "feel good" being recognised and are more likely to view your presentation favourably. You need to accumulate goodwill early in your talk. Some sponsors have great name recognition, which indirectly enhances your credibility. If the sponsor has a logo, use the logo instead of the name if you are short of space. Presenters often acknowledge people and sponsors at the end of their presentation, after the conclusion slide. It is not the best time to do it, because you may be out of time and may have to skip the acknowledgments. Finishing your presentation with the acknowledgment slide prevents you from entering the Q&A session while your conclusion slide is still on the screen, helping the audience remember the main points of your presentation.

*Do I need to restate my name and the title of my presentation
since it is already on the slide?*

That depends. If the chairperson has introduced your name and the title of your presentation, do not repeat. Otherwise, you should certainly give your name. Do not rush through this. Of course, you are very familiar with your name, but other people are not. Therefore, articulate it slowly. Give a little comment on it, if it is original. It makes you feel more human. For example, I often start my classes by writing my name on the whiteboard and explaining that hyphenated first names starting with "Jean" (in my case "Jean-Luc") are very frequent in France. I then give another ten examples as fast as I can while smiling. I am still "breaking the ice", warming the audience to my personality. I do not recommend using the turnkey formula I hear everywhere: *Good morning, my name is xxxx and the title of my talk this* [*morning, afternoon, evening*] *is xxxx.*

The title of your talk is on the screen. People can read it. Explain it. Clarify it for the non-expert audience. Turn it into a question, or paraphrase it using simpler words. However, for those of you who have a strong foreign accent, it helps to read the title slowly so that people can get used to your accent and associate what they read with what they hear. It makes it easier for them to recognise these words when they hear them again later.

Do I really need to write the many names of the co-authors of my paper?

Yes. Recognise them. Use abbreviations for the first names, and keep every name equal in size. Just make clear what your name is among the other names, either in what you say, *I am the first name of this list*, or by discretely changing the font colour (black for you, grey for the others — avoid red!)

My title is long. Can I change my title and write a shorter one?

The title of a presentation creates expectations. If your title is different from that printed in the program, people may have different expectations or may be confused. You could start with the initial title and then through software, make it fade out and replace it with a shorter and clearer title if you have one, as long as this newer title raises the same expectations as the initial one of course.

I am not the first author of the paper presented. Should I tell the audience?

Yes. Unless you know as much as the first author, it is best to set the expectations of the audience early so that, come Q&A time, people do not ask you the questions only a first author can answer. That is why it is better to decline invitations to be a substitute presenter if you are not as knowledgeable as the one you replace. Your reputation is at risk. Do not be lured by the opportunities to go to exotic places and stay in fancy hotels. Believe me, a bad question-and-answer session is enough to destroy your chances of hiring, being hired, being funded, or being sought out for research partnerships.

Hook Slide

What?

- Surprising result
- Problem, question
- Reason for research

Where?

- After title slide
- On title slide

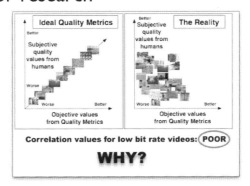

Hook Slide — The Attention Grabber

A — The hook slide focuses the attention on the presentation theme

1) It helps focus audience attention on the scientific contribution. Initially, the audience may be unfocussed, distracted by late arrivals or chatty neighbours. Have you noticed how long it takes you to get into a talk when you are part of the audience? There is a moment of inertia before your brain starts paying attention regardless of how much scientific interest the title of the presentation created. The airplane is still taxiing on the runway; it has insufficient speed to take off. During that time, the audience is in discovery mode, focused on your personality, but not quite yet on your scientific topic. It would therefore be wrong to hurry head first into a topic without giving time to the audience to warm up and without first

channelling attention to the scientific topic. The hook slide, while keeping the attention on you, moves the audience to the content discovery stage.

B — *The hook slide raises interest high enough to keep floaters in their seat*

2) At the first inkling that the talk is going to be uninteresting, floaters usually get up and leave. They came to your talk out of curiosity hoping to garner some useful nuggets of knowledge. They often lack the background that experts have and are not as motivated as those with a set objective who intend to stay. Your first objective is to convince the floaters that they will understand and benefit from your talk. They need to be hooked to your topic (and nailed to their seat).

C — *The hook slide gives you the opportunity to express how interesting your research is*

3) Beginnings of talks are dull because the first slides are dull. They do not allow you to be energised by your topic, and to energise the public at the same time. How can anyone be energised by an outline slide, or a text-heavy motivation slide? A good energising hook slide relies on visuals, animation, or video to garner interest.

4) The contents of the hook slide garner only part of the interest. Human interest is contagious. Interest is expressed through a higher-pitched louder voice, dynamism of gestures, and brightness of the eyes. The audience can detect your interest as surely as it can detect your boredom.

D — *The hook slide creates the expectation of an enjoyable presentation*

5) Any great start creates momentum. It carries the audience forward into the rest of the presentation. Remember those movies or theatre plays that start in action and how quickly and easily you followed and were absorbed in the story? The hook slide enhances the appeal of the story slides. If your hook slide is effective, the audience will want to stay to hear what you have to say.

Frequently Asked Questions

Does the hook slide play the same role as the introduction in a scientific paper?

In a scientific paper, the introduction sets the theme and scope of the research, and it answers the *why* questions: why this, why this way, why now, and why should the reader care. In a movie or play, the director also needs to introduce the main characters (*who*), the parts that will ultimately create a problem (*what*, *why*, and *how*), and the context (*where* and *when*). The hook slide is usually at the same place as the introduction of a paper: after its title; therefore it would be sensible to argue that it should play the same role as the introduction, and be built using only material found in the introduction.

You face the same problem a movie director faces. Both of you do not have much time to spend on the introduction. The director, therefore, starts his movie in action, pulling in the spectators into the core of the story as early as possible, stereotyping the characters or focusing only on a few character traits that are revealed in action. In scientific presentations, putting the hook slide right before the title is risky however. It grabs the attention of the audience so much that it eclipses the title slide. When in that position, the hook slide should be presented briefly and time should be spent explaining the title slide and how your title relates to the hook slide.

Introduction comes from the Latin "*ducere*": to draw. The introduction draws the audience into your topic. As I typed this, I misspelled *introduction* and, instead of "*d*", typed the letter to the left of *d*, which happens to be the letter *s* (on a qwerty keyboard). The hook slide is more an *introsuction* than an introduction. It does not draw the audience into your talk, it sucks it into your talk in less than a couple of minutes. Suction creates a vacuum often used to create adhesion. The hook slide should do just that. It should create the aspiration to know more while also making the audience adhere to what you say (acceptance through belief). You cannot afford to lose believability early in your presentation.

After the title slide, I often see a text slide with the heading "motivation". Is that the hook slide?

If your motivation is self-centred, your hook may fail. The reason you did your research may leave the audience cold. In other words, just answering the question *Why am I working on this* is not enough, even though it is the reason why you decided to spend months of your life pursuing (and hopefully cornering) scientific truth. But if you answer the question *Why should you, the audience, care about the fruit of my research*, stating your motivation would be a great way to hook the audience ... if expressed visually. Too many motivation slides lack appeal, not because the motivation is wrong, but because it is given to the audience to read in paragraph form.

In conclusion, hooking by illustrating your motivations is a good way to start, but it is only one of many ways. Consider the other ways described later before you make your choice.

What makes a bad hook slide?

I compiled the audience feedback on how scientists perceived the hook slides from 167 presentations. Approximately four out of five (131) were evaluated as having an unsatisfactory hook. I then classified the reasons given by the participants into four classes of bad hooks.

1) The hooks loosely or remotely related to the title, even if visually interesting

These hooks are bad because they raise false expectations. I remember a slide with a great shot of surgeons in green gowns in an operating theatre surrounding a patient whose midsection was cut open. Soon, in large bright red letters overlaying the picture came the question *Could you be your own donor?*. The audience expectation raised was that the presenter would answer that question before the end of the talk. The question was never answered because the talk was about engineering scaffolding for tissues. Yes, one could provide one's own organ tissue to be grown on these scaffoldings ... but that was not the focus of the presentation. I also remember a young scientist who started his talk by asking a personal and direct question to the audience, expecting someone to respond. *Who among you has a relative who died from cancer?* he asked. That question was far from his topic: bacteria suspected to be responsible for stomach cancer. No hand was raised, not even a finger. The question was too personal, and the audience still cold. After an embarrassing silence, he had to respond himself, became nervous, withdrew from the unresponsive audience, and did not do well subsequently.

2) The excessively long hooks eating into the core of the presentation time

The airplane remains taxiing on the runway for far too long. By the time it finally takes off, the plane does not have enough fuel left to complete the journey, and has to go through a crash landing. This bad hook is often visually interesting but then the presenter gets stuck on the hook slide(s) either because the diagram used to hook is so complex that it requires far too much time to explain, or because the presenter gives the audience a far too substantial background. The hook is nothing but the appetiser to the scientific meal. It stimulates the appetite; it is not the main dish. How would your guests feel if you offered them bowl after bowl of salted peanuts, and then rushed through the main course and cleared the table while people were still eating?

3) The text-heavy hooks

A word-only slide is rarely an effective hook slide. "Motivation" or "Objective" text slides detailing in paragraph or bullet form what the presenter intends to demonstrate make bad hooks. Do not get me wrong, making the objective clear to the audience is laudable, but it should not be done through a text-heavy slide that puts the audience in "reading mode" right from the start of the presentation. Such a slide immediately sets the expectation that the presentation is going to be text-based, i.e. screen-based, not presenter based. The message given by such slides is "*watch the screen and read*". Do not pay attention to me, the presenter: I am not essential; everything is on my slides.

4) The highly technical hooks for experts only

Starting too technical is just as bad as starting too slowly or from too remote an angle. The non-experts need context or knowledge to follow your points. In the original movie "*Three men and a baby*", a bachelor has a baby dumped on his lap without notice and without baby supplies. The unprepared bachelor goes to the shop to buy a bottle. The shop attendant asks him which teat size he requires. The bachelor does not have the knowledge to make that decision (the older the baby, the larger the teat, because an older baby can deal with a greater flow of milk without choking). Likewise, if you want to hook your audience, it has to have the background necessary to understand your hook.

5) The over-promising hooks belying the conclusions

A hook cannot be artificially hyped to attract attention because it raises expectations that the rest of the presentation will not able to fulfil, thus creating a gap and frustrating the audience.

What is a good hook slide?

A good hook slide is one that very quickly captures the attention and sustains it after the slide is viewed. It makes the audience want to know more. I observed that people using the hook slide to state the research problem were more successful than others. It creates a better vacuum. It raises the questions *why this?* and *why should I care?* I also observed that good hook slides were visually striking. They relied on images, animations or movies. When text-based, they often contained one short question or statement written in extra large font size. Aside from explaining the motivation, good hook slides excel at raising questions. That is their hallmark. The answer to these questions is given in the story slides that follow the hook slide. I have seen good hook slides state a striking result (the contribution), and the rest of the presentation simply answered the question: *How did you do this?* This last method is quite effective because it allows you to deliver the take-away message right up front in your presentation. People tend to remember what they hear first and last, not so well what is in-between.

Can there be more than one hook slide?

Remember the reason for the hook slide: to quickly capture the attention and sustain it. As long as you keep that in mind, you can have more than one hook slide, but keep the slides moving — do not linger on them. In a longer presentation (45 minutes or more), there may be several hook slides, one for each main section of your talk.

My presentation topic is really highly specialised and I suspect that the audience requires some background to understand. Do I provide the background before or after the hook slide?

Firstly, please accept my heartfelt thanks. You are thinking audience. This is excellent. In addition, you raise a very good question, one with no easy answer. In a way, you know you need to provide background, but if you do it after the hook slide, it destroys momentum, and if you do it before the hook slide, it dulls your start. You are in a difficult situation. To provide background in an interesting way, explain new words or concepts in a just-in-time fashion, starting with the title slide. The key here is brevity. You cannot spend time educating, but you can clarify using words or metaphors your audience understands. Explaining just in time has two advantages: you keep the story moving, and your audience does not have to remember things you told them earlier.

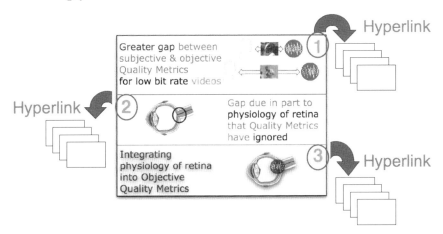

Watch the "Map Slide" DVD Movie to discover how to make a map slide using PowerPoint or Keynote

Map Slide — The Head's Up Option

A — The map slide maps out the rest of the presentation to help the audience keep track of time

1) In a short scientific talk, the story is usually easy to follow since the number and complexity of arguments are limited. In a longer talk (30 minutes or more), more is presented, and often with less urgency. In long talks, the

audience's attention is likely to have more ups and downs than during a short presentation, and unless one is a master storyteller, fatigue may make some restless enough to want to know what is left to present. Without an outline, the audience is kept in the dark, a frustrating situation. The outline should be brief enough or be logical enough to be remembered because the audience mentally checks off the points that have been covered in the memorised outline. It is appropriate for you to refresh people's memory by stating what is yet to be covered in the presentation. Long or disorganised structures are impossible to remember unless one remembers the logic that connects their parts.

B — The map slide reveals the logical structure and makes clear the process followed to reach the conclusions

2) The map slide plays the role of the *informative* headings or sub-headings in a scientific paper. Informative is italicised because in a scientific paper, not all headings are informative. Standard headings such as introduction, methodology, or results are only section bookmarks unrelated to the title of the paper. Such uninformative outlines have no place in your map slide — or any other slide for that matter. Informative outlines reveal the structure of your storyline, and how it relates to the title.

Frequently Asked Questions

Why is the map slide optional?

If your presentation is short (less than 30 minutes), orientation into the rest of your talk does not require a slide. Instead, orient the audience orally after presenting your hook. For a short presentation, there is no need to reveal the structure. In printed scientific communications, the short paper or letter does not require headings; only long papers do. There are far too many boring outline slides in scientific meetings today. You may go straight from the hook slide into your story. If the story is interesting, the audience does not feel time passing, nor does it become impatient and wish to know what comes next in your talk.

If your presentation is long (30 minutes or more), the map slide is useful because the audience is more likely to become impatient and wants to know where you are going with your reasoning, particularly if that reasoning involves many steps.

Is the map slide an outline of the paper structure or an abbreviated abstract?

If you view your map slide as an abstract in line with the typical structure of a paper: introduction, methodology, results, and discussions, you are likely to fall into a common trap. This standard structure encourages multi-point slides with vacuous headings such as "background", background (ctn'd or continued), "results (part 1)", "results (part 2)", "additional results", "further results". It is a cage that works for papers but not necessarily for oral presentations. Enjoy the freedom to open the cage, a freedom only found in oral presentations.

If you view the map slide as the successive headings of your talk (not of your paper), it becomes a guide that helps the audience keep track of time while allowing you to structure your talk to better capture the interest of a live audience. The order followed to demonstrate the validity of your contribution need not be as rigid as the well-ordered steps of the scientific process itself. Let the story control the step order. You may skip steps: for example if your methodology is nothing new to the audience, a verbal mention suffices. You may coalesce steps: self-contained slides combine result and interpretation. Or you may even put the results before the method if doing so improves the story (for example when your method is your contribution).

Shouldn't the map slide come before the hook slide?

This view is derived from the standard structure of a written paper where the abstract comes before the introduction, i.e. the map slide comes before the hook slide. We have already identified that a scientific talk is not a written paper, and therefore does not necessarily need to adopt the structure of a written paper. In scientific talks, new considerations come into play such as the need to attract the attention early and to focus that attention on you first. Such considerations are absent in a written paper. It is difficult to engage the audience with a map slide. Remember that the way to engage the audience does not only rely on content, but also on your energy, dynamism and demonstrated interest in your topic. A "table of contents" type of slide is as exciting as a shopping list.

My presentation is one hour long; what does the map slide look like?

The best map slide is visual. Let us say that your presentation had four different sections (three would be better). Your map slide would have four visuals, each with redundant explanatory text. Each visual would be hyperlinked to a part of your presentation. After each part is completed, you would return to the map slide and navigate (again using hyperlink) to the next section. It is that simple. In addition, the map slide has the advantage of letting the audience keep track of time while reinforcing the structure of your talk by repeatedly showing its main points briefly. The conclusion slide follows the end of the last section or is reached by clicking on a hidden button located on the map slide. By the way, I do recommend that you structure each section as you would the whole presentation (recursively). Each section would thus have a hook, a story, and a brief conclusion. This recursive model enables you to capture and sustain the interest longer.

Once my outline is presented, do I need to refer to it again?

In long presentations, referring (at least orally) to the visual outline points allows you 1) to pause and consolidate knowledge by reminding the audience what has been accomplished so far, and 2) to introduce the next point in the outline, and the next slide in the talk.

In short presentations, if you do a spoken outline after your hook slide, do so while blanking the screen, thus keeping the audience focused on you, not on the screen.

Can my outline be nested (a top level outline and an indented outline for each sub-level)?

If your presentation is several hours long, and the audience is ready for it, you may want to do that. However, nested outlines are not recommended for normal presentations. If people use nested outlines, it is to mimic the structure of a published paper with its headings and sub-headings. They are fine in a paper, but not in a scientific talk. They demoralise the audience which fears presentations with long lists of nested bulleted items. Lists are not only difficult to memorise, they also fail to establish the relative importance of one item in the list versus another, and they fail to reveal the relationships between the items. A graphic or visual outline is more effective in that regard.

 **Listen to the "Core competitive advantage" podcast on the DVD**

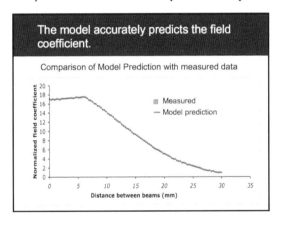

Story Slides — The Proven Claims

A — Story slides fulfil the expectations created by the title and hook slides

1) The story slides ensure continuity of purpose. You set up the stage with the title and hook slides. The audience is now waiting for you to deliver on your promises. Answer the questions you raised. There should be no disconnect.

B — Story slides succinctly present and justify the novelty claims

2) The claims are part of the novelty you highlighted in your paper. Not all novelties, however, are included in your presentation, only those that participate to the coherent, believable story announced in your title.

3) In keeping with the scientific process and the principle of self-contained slides, each text-based claim made on a slide is immediately (on the same slide) justified by visual/data proof.

4) Each claim is written succinctly and justified by the most convincing argument. If a claim remains convincing after a detail is taken away, then that detail is not essential to the story. Keep things moving during the formal presentation. There is no need to crowd the slide with secondary proof. It can be mentioned orally only. Later, during the Q&A, if the question arises, provide more detail and more complex visual proof.

5) If adding or removing a claim or piece of evidence does not clarify the story, I would argue that either that claim or piece of evidence is not essential to the story, or it is misplaced and needs to be moved to its logical place.

C — Story slides present the claims in sequence to form the tightly knit and logical story announced in the title and hook slides

6) Your story is a bird's eye view on your helpful scientific contribution. In this view, all things are logically linked. If you cannot think of what to say to establish a verbal transition from one slide to the next — if the only words that come to your lips are *and here, on this slide* — then let me argue that your story is not logical. You need to rework your story plot.

7) The story is tightly knit. If you can move a slide to any position in your slide show, then that slide has no place in your presentation. If you can move a slide to two different positions in your slide show, then either that slide, or the slide before or after that slide should be removed from your presentation.

Frequently Asked Questions

The format recommended for the normal slide is quite unusual. Where does it come from?

I discovered the format about five years ago through the Virginia tech website of Professor Michael Alley. I was sceptical at first, but after having seen it used successfully by hundreds of researchers in A*STAR Research Institutes over the years, I am convinced of its worth and have become an advocate. It is better than the "classic" way of building sides, i.e. headings that act as signposts and are low in information content, horizontal rakes of bulleted items that make more than one point per slide, and low resolution visuals that are cut and pasted versions of high resolution PDF files. Naturally, this new format requires more thinking, more story planning, and more focus on the audience. If you want to know more about this new slide design, visit the following website:

 http://www.writing.engr.psu.edu/slides.html

Professor Michael Alley refers to the design as "the assertion-evidence structure".

What makes a bad story?

There are many bad stories.

1) A bad story is one where the audience cannot pick up the story thread as you move from one slide to the next

The audience fails to see the transition logic for three reasons:

A) The audience does not know what to expect next in your slides because you have not set any expectations; your claims are simply juxtaposed. Even if you changed their order, it would not matter because they are only loosely related. Indeed, the way you introduce such slides is often with the word *next*.

B) The audience is lost because your slide created an expectation of what comes next, but you disregarded this and in the next slide moved to a completely unexpected new topic.
C) Your transition is logical, but to you only, and you have not explicitly revealed the transition logic. The audience does not have your expertise and therefore fails to see the logic in your transition. Just as in a written paper, writing the words *it follows* does not necessarily mean that the reader follows at all. In a scientific talk, saying *as a result* or *therefore* does not mean that the audience clearly sees the cause and effect relationship. Your help is required.

2) A bad story is too complex to follow

It usually taxes the memory and relies excessively on information revealed in past slides. Complexity is born out of the abundance of unexplained details, acronyms, and tables of numbers. Complexity is also born from a lack of imagery to explain simply what is complex. Finally, complexity is the result of a presenter's unwillingness to bring down word complexity through a more accessible and a better explained vocabulary.

In this bad story the pace is often rapid, and the oral narrative accompanying the slides is often sketchy while the written narrative is quite the opposite, as if the presenter expected the audience to learn from reading the slides only.

3) A bad story is one where a slide makes multiple claims

The one-slide/multiple claims situation is frequently found because of the ease of adding a bulleted item to a list. Bulleted text claims are made without their essential supportive visual evidence. Despite the accumulation of claims, believability is weakened. Here again, more ends up meaning less. Such claims are often questioned by the audience during the Q&A leading to a possible loss of credibility. Do not show what you are not prepared to justify.

4) Finally, a bad story is one that is too simple, too general, or too superficial

After hearing it, the audience is disappointed because nothing useful came out of the talk, just generalities. People start doubting your expertise because experts go into details; non-experts remain vague or general. Keeping a presentation clear and accessible to the majority of your audience does not relieve you from the duty to provide precise technical evidence for specific claims.

What makes a good story?

A good story is the result of much work. It starts with sorting out all the novel and useful material from your paper, finding a good angle to present it (the hook), and then selecting a subset of this prime material that can be woven into an interesting story plot. There are many such subsets, and indeed, there are many possible story plots, not all of them interesting. As long as you understand that you do not have to present ALL the new and useful material, but just the most interesting material (i.e. as long as you think audience), your story will probably be good. A good story has its parts well connected by spoken transitions. There is always a connection between two slides; therefore, there is always a need to make clear to the audience what the connection is through spoken transitions.

What makes a good slide transition?

In film making, transitions come in a number of forms:

A) A dissolve from one scene to another
B) A subject shown at contrasting times such as when a scene ends on a close-up and the next scene starts on the same close up but after days or years have passed
C) An audio/video split where the video of the new scene arrives before the audio from the old scene finishes, or when the audio from the new scene arrives before the video from the old scene finishes
D) A cut from one scene to another (jump cut in time or space)
E) A fade to black and fade from black, a fade to white and fade from white
F) An event triggering a scene change (phone call, departure, anything that can only be fulfilled outside the current time/place)

As you see, methods to establish transitions are abundant; some rely on the story plot itself, others on video editing techniques.

Transition methods are just as abundant in scientific presentations. Constantly hearing *And on this slide*, or *And next*, or *And here* to transition from one slide to the next is similar to watching a movie where all scenes are abrupt jump cuts. Film director Jean-Luc Goddard enjoys fragmenting and deconstructing a film through editing. He forces the audience to be attentive by disclosing the whole time plot as pieces of a time-independent jigsaw. Unless you are a Jean-Luc Goddard, use one of the well-proven transitions techniques described hereafter.

1) Transitions based on logic

These transitions are based on the logic of an argument. For example, the presenter demonstrates that logically, to prove E, one would need to prove that A is true, B is false, and that C and D are neutral. The order does not really matter here, but all facts have to be examined one by one. The presenter would first present the logic of the argument and convince the audience that the logic is sound before covering the logical steps leading to conclusion E.

2) Transitions based on sequential process

Such transitions are based on the need to move from one step to another. The sequence of steps has been announced at the beginning of the sequence and each slide corresponds to one step in the sequence. The sequence could be based on time, or based on the scientific process (from observation to theory, from theory to experiment, and from experiment to theory validation).

3) Transitions based on fade to black

In movies, directors use the fade to black technique when they need to move the audience from one topic to another. In a scientific book, it would be from one chapter to another. The fade to back is an outstanding technique to get an audience easily hypnotised by the large bright screen to refocus on the presenter. This technique is so necessary that all publishers of presentation software have embedded a "cut to black" functionality in their software, and even manufacturers of presentation remotes have a button to cut to black. It is called the "B" key. Within a slideshow, if you press the letter "B" on the keyboard, your slide disappears and the screen behind you goes blank, as if you had no projector at all. As soon as you press the "B" key again the same slide reappears on the screen.

The fade to black transition is appropriate when you need to pause to summarise your findings prior to introducing the next stage. But it could also be used to give your audience a rest while you tell a personal story that relates to what has been shown so far.

4) Transitions based on the need to change scene

These transitions usually rely on a question and ask the listener to think of an answer prior to providing it in the heading of the next slide. Imagine that your current slide displays a porous material that carries enzymes for catalysis, but whose leaching is high. You would transition by saying *As you see, the pores on this carrier are of irregular sizes. This favours leaching. Is there a way to engineer a carrier whose pore diameter can be controlled?* The next slide would present a new type of mesocellular foam.

5) Transitions based on a graphical map slide with hyperlinked visuals

When the presentation lasts more than 30 minutes, it is helpful to display the story plot in a visual map slide revealing the three or four sections of your talk. When one section is completed, you return to this visual outline slide where each section is hyperlinked to its corresponding set of slides. This map slide serves as a transition device into the new various sections. Some presenters use this type of transition for normal story slides. They jump back and forth between a story slide and other story slides via hyperlinked graphical or textual objects. It has been my experience that audiences do not like such transitions. They prefer smooth continuous transitions that are story-based, and not hypertext-based.

Is it good to use PowerPoint or Keynote slide transitions to attract the attention of the audience?

Slide transitions are more distracting than helpful. They are gimmicky and typically used by people who start using PowerPoint for the first time (secondary school level). Do not expose yourself to ridicule or risk looking amateurish. In scientific presentations, avoid slide transitions. However, I can think of a few situations where they are useful:

1) To replace a "B" key (*fade-through-black* slide transition)

2) To emphasise an oral transition statement that has an equivalent in a visual transition as in "*When you look down toward the ship's keel, you see...*" and as you say these words, you use a slow *push up* effect that gives the impression you are taking the elevator (lift) down.

3) To enhance the message that you are going to a different section of your presentation on the map slide (*cube* effect or *barn door* transitions for example).

4) To demonstrate continuity of a document spread over several screens (*push* effects)

Watch the "Moving through a large document" video on the DVD

5) To simulate the effect of builds using multiple slides instead of one. For each new element introduced or removed, a new slide is created. That slide is a duplicate of the previous slide. Thus, you can fade the new elements in or out of the slide by using a *wipe* or a *fade-smoothly* slide transition.

What does a typical story slide look like?

The title in the title bar states the claim in text form (usually as a well-formed complete sentence with verb). Under the title bar, a visual provides the evidence to substantiate the claim.

What process do you recommend to put the story together?

If you are someone who likes pen and paper, take one A4 sheet of paper per slide, turn it horizontally to represent a slide on your screen, and write one claim (one only) at the head of each page. Then shuffle and position the pages so that they are appropriately ordered, which means that you have no difficulty finding a transition between a slide and the next. Use a white page to represent your "B" key. Once you are pleased with the four or five headlines (for a seven-minute presentation) because they are interesting and well hinged to the title and hook slides, then you are ready to look for the visual evidence that will substantiate each of your slide headlines. If you like to work directly on the computer, create enough blank slides and type one story headline in each. Then go into the slide sorter view (PowerPoint) or light table view (Keynote), and rearrange the slides in perfect order by dragging the slides to their right place, hiding or deleting the ones you feel unnecessary, and reworking the wording of the title of each slide so that it fits nicely on one or two lines.

Should each story slide display the company logo?

If your research organisation imposes the use of a slide template, it will certainly put its logo on the slide master. Use it on your story slide if you wish, but whenever the logo gets in your way, be ruthless and remove it. For this, right click (Mac: option-click) on an empty part of your slide background, select "*slide background…*" and select the checkbox "*Omit slide background from master*". For Keynote users, change the theme or pick another Master slide in the theme.

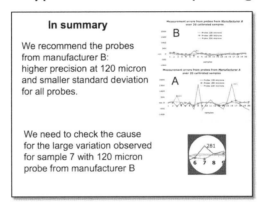

Conclusion Slide — The Promised Items

A — The conclusion slide repositions the claims as take-away points for the audience

1) In a scientific paper, the author constantly repeats the contribution. As the saying goes, "tell them what you're going to tell them (your outline/hook),

and then tell them (your story slides) and tell them again what you told them (your conclusion)". But instead of a pure restatement of each claim, invite the audience to consider the possible benefits or applications your proven claims enable you to suggest, or even call the audience to action on the basis of the evidence provided (need for further research, for example).

2) Note that nothing is ever new or unexpected in a conclusion. Everything is predictable. If you wish to test the anger of an audience of scientists, make sure to insert unproven conclusions or unsubstantiated suggestions on your last slide and wait until the Q&A for the demolition work to start.

B — The conclusion slide ties back to the title and hook slides

3) The title and hook slides make many promises and raise many expectations. The conclusion slide shows that these promises have been kept and the expectations have been met.

4) If the title and hook slides hint at the usefulness of the contribution to the scientific world, the conclusion slide has to keep that audience-centric point of view in its wording. I remember a presentation where the hook slide whetted the appetite of the audience by raising the possible role of a gene in a rare stomach cancer leading to possible therapeutic strategies. The conclusion slide, however, mentioned the gene but remained silent on the therapeutic strategies. There was a clear gap between the expectations raised and the expectations fulfilled.

C — The conclusion slide allows the audience to ask questions more easily

5) The conclusion slide, the last slide of your formal presentation, remains on the screen during the Q&A to enable the audience to remember the main claims and their corresponding supportive visuals. Remembering a whole talk is not easy. If your last slide is the acknowledgment slide, or an ominous "thank you" slide, or a "Q&A" slide, or worse a black slide, you are forcing the audience to recall the content of your talk without any outside help. It is little wonder, that so few questions are asked when the last slide leaves the audience literally in the dark.

*D — The conclusion slide allows you to make your own
purpose clear*

6) Information downloading is not the reason for your talk, and information uploading is not the reason why the audience came to listen to you. Therefore, now is the time not only to rephrase your contribution but also to make your purpose clear. For example, if your purpose is to solicit research cooperation, your whole talk should be structured to make clear the unresolved research issues. Your conclusion slide would then remind the audience of these, and your accompanying words would offer partnering opportunities with others working in the same field.

Frequently Asked Questions

How do I transition into my conclusion slide?

To transition between the last story slide and your conclusion slide, while still on the last story slide, or on blank screen (from "B" key, or fade to black), announce that the time has come to conclude your presentation. If you stand on the side of the podium far away from your computer, walk back to your computer while the screen is blank. Smile and remind the audience of what you promised at the beginning of your talk, then bring in the conclusion slide and, point by point, show that you delivered on your promises. This explicit link to the start of your talk brings good closure.

*How can I possibly fit the main text claims AND the visuals
on the conclusion slide?*

There is not enough space for the visuals to be readable, but fear not. These visuals are scaled-down jpeg representations of the originals. They are not meant to be readable, just recognisable. As to the text claims, they no longer need to be whole sentences: bulleted phrases are acceptable here. Keeping the text close to its corresponding visual reminds the audience which visual supported which claim/suggestion/call of action. Each miniature visual is hyperlinked to the full scale visual, thus enabling you to jump directly to a slide in answer to questions.

There is only ONE conclusion slide. Therefore, be selective. It is not necessary to feature all the points made during your talk. Some of these points are just stepping stones towards the larger points. Select those that are fulfilling immediate needs, yours and the audience's.

When I jump away from the conclusion slide during the Q&A, do I need to return to it?

It all depends. Very often, once you go to a story slide to answer a question, the next question from the audience will target the same slide. If you do not mind, then there is no need to return to the conclusion slide. If the story slide you landed on is a controversial slide, or a troublesome slide in terms of questions, move back to the conclusion slide before answering other questions. You can do so easily. Just take note of the number of your conclusion slide prior to your talk. Let's say it is slide 14. To return to slide 14 from any slide in your slideshow, just go to your keyboard and type "*1*" "*4*" "*enter*", and you will be returned to the conclusion slide.

Do I talk about limitations or future work while on the conclusion slide?

Unless your aim is to establish research collaborations, there is no need to return to the limitations you mentioned while presenting your story slides earlier in your talk. Stay positive at the end of your talk. However, be sure to mention the limitations of your findings during your talk before the Q&A session, to pre-empt questions aimed at discrediting your work in front of the audience by minimising its usefulness.

As for future work, unless your goal is to secure collaboration for research continuation, keep that as the question you ask to yourself (or the question prepared for the chairperson) if the audience has no question for you at the start of the Q&A.

Do I write "thank you for your attention" at the end of the conclusion slide?

NO, NO, NO. It is a crying shame to see a computer screen thank the audience when the presenter is live in front of the audience. Nor do you write "time for questions" or any other reference to the Q&A on the conclusion slide.

This is something for you to say. You are the one answering questions, not the computer.

Do I write the heading "Conclusions" at the top of my conclusion slide?

Since you announce the conclusion before bringing the conclusion slide to the screen, it may not be necessary to do so; furthermore, if your conclusion slide is crowded, you may want to remove the "*conclusions*" or "*summary*" header, which occupies a large chunk of prime space at the top of your screen. On the other hand, if your presentation is short and your conclusions are few, there is no harm in writing "*conclusions*" or "*summary*" at the top of that slide.

Is the conclusion slide the last slide?

The conclusion is the last slide presented before the Q&A but not the last slide of your slide show. Many slides follow the conclusion slide. These are the slides you have prepared to answer more technical questions, should they be asked during your Q&A. Having slides after the conclusion slide also prevents situations where the nervous presenter clicks past the conclusion slide (the last slide) and quits PowerPoint before the Q&A, thus having to restart it, retrieve the conclusion slide and display it again.

If the conclusion slide is your last slide and if you use Microsoft PowerPoint, open the preference panel, and select the option *End with black slide* in the *view* preferences. Returning to the conclusion slide from the black slide is easily done by pressing the left arrow on your keyboard or on your presentation remote. If you use Apple Keynote, open the preference panel and under the slideshow preferences unselect the button *Exit presentation after last slide*. The last slide will remain on the screen until you quit the application or press the escape key.

How many points can I make on my conclusion slide?

The points made on a conclusion slide are the points you want the audience to remember, the take-away. Therefore, focus on only two or three. The conclusion slide is not a summary of your talk. It is the final word on the strength and impact of your contribution. The greater the number of points, the more likely the audience will forget the most important ones. Within three days, most of your audience will have forgotten the majority of your points. This may sound depressing, but it is a fact. Therefore, here again, less is more. Express your conclusion points in few words, thus avoiding long unmemorable sentences.

6

SLIDE DESIGN

Design for Slide Legibility

In a presentation setting, legible text is text that can easily be read by a person sitting at the back of the presentation room. Legibility at a distance is what matters. The audience has the right to seat wherever it pleases, so asking the folks sitting in the back row to move forward under the pretext that you have a lot of small text on your slides is not an option. Legibility at a distance is not simply a matter of ensuring that the information on the screen is legible if people squint their eyes. As Edward Tufte, the visualisation guru puts it: "*we should never be working at the edge of legibility in any situation*". Everything on a slide should be legible without effort.

Text legibility is not just a matter of selecting a font with the right size. There is so much more to it than that. Experts in presentation skills offer their own recipes for legibility. Problem is, they disagree with one another: *use only 6 words per line and only 6 lines, … use only 5 words per line and a maximum of 4 lines … use only font sizes equal or greater than 24 points … never go below 18 points in plain or 16 points in bold … do not use a white background … use a white background … use sans serif fonts only … use any font you like, serif or sans serif …* and so on. Who are we to believe? Everyone is right, and everyone is wrong. There is no shortcut, no one size fits all, no magic formula. Legibility at a distance is not easily determined because it is the result of the interplay between document design, environment, and technology.

Document Design

- The text factor

  Font type
  Font size

Font style
Slide dimensions
Spacing between characters
Spacing between words
Spacing between lines
Length of lines
Orientation of text: vertical, horizontal, and diagonal
Density of text
Alignment of text
Spelling and grammar: error free or mistakes
Vocabulary/language: known or unknown

- The colour factor

  Colour of text foreground
  Colour of slide background
  Colours in image

- The image factor

  Resolution of image
  Size of image
  Contrast and brightness of image
  Thickness of lines in image

- The layout factor

  Ordering of information on slide
  Spacing between elements

Environment

- The meeting room factor

  Light level in meeting room
  Light level on screen
  Size of projection screen
  Placement of lectern versus screen position
  Distance from back of room to screen
  Obstacles in path of screen

- The audience factor

  Age and eyesight of people in audience
  Placement of people in room
  Culture of people in room

Technology

- The data projector factor

    Contrast radio 1:1500 or 1:2000
    Resolution: XGA or SVGA
    Offset of projection beam with screen centre
    Keystoning correction: compression or scaling
    Technology: Polysilicon, amorphous silicon, LCD, DLP
    Luminosity and remaining life of Metal Halide bulb
    Brightness and black level
    Misalignment of projector

- The presentation software

    Microsoft PowerPoint, Apple Keynote, Adobe PDF, Macromedia Director, Omnigraffle, Novamind and others

Listen to the "Pearls of presenter wisdom" podcast on the DVD

17 Questions on Slide Legibility

Question 1: It seems my slides are more readable in one meeting room than in another. Why is that? Does that have anything to do with the projector brand?

Forgive the pun, but there is more to legibility than meets the eye. A legible font type and size is impossible to decide in isolation because it is influenced by meeting room size, screen size, room ambient light level, distance of the audience to screen, age of audience, and projector resolution and brightness (number of ANSI lumen). To best understand the relationships between projector, screen, and audience factors, it is useful to play with a projection calculator. An excellent one is found here:

 http://www.projectorcentral.com/projection-calculator-pro.cfm

The "what-if" of projector calculator simulation, however, does not show what the projected image looks like at the settings chosen. **Projectors with a short throw** (they throw a bright wide beam at a short distance from the screen) have specific advantages over others: they allow the presenter to be close to the screen without being blinded by the projector light. With such projectors, the image is bright, the text more legible, and there is no need to dim the meeting room lights, enabling the presenter to keep eye contact with the audience. Unfortunately, the projected image is trapezoidal, not rectangular (keystone distortion). After correction, projected text loses clarity, forcing you to increase font size to keep text legible.

Projectors with a longer lens that are aligned with the screen centre and that have the same number of ANSI Lumen as the short throw projectors project an undistorted image. Unfortunately, the image is less bright because the projector's long lenses absorb more light. Nothing is perfect. Trade-offs are everywhere. After this brief discussion, you may think that brighter is better. Alas, even a modest 1400 ANSI projector can be too bright in a small meeting room, resulting in headaches and eyestrain. To prevent such pains, some projectors have built-in low light settings and brightness level control, but not all.

As presenter, you have little influence on the type of screen and projector available. You also have little influence on meeting room choice, screen size, and distance of audience to screen. But it is your responsibility to keep yourself visible and your text legible. What can you do? Two parameters remain under your control: the meeting room lights, and your slides.

Dimming the meeting room lights is not the answer even if it does make the fonts on the screen more legible at a distance. Dimming lights turns you into a ghost lurking in the shadows of the lectern. Dimming lights also sends your audience to sleep, and dilates the eye pupil thus reducing eye resolution and ability to see contrast. For the science behind these facts, read the comments of Niels and Alex Merz on Edward Tufte's website discussion on "recommended background for projected presentations".

Design for Slide Legibility 101

 http://www.edwardtufte.com/bboard/q-and-a-fetch-msg?msg_id=000082

Now that you have ruled out dimming the meeting room lights, one factor, and one only remains under your control: YOUR SLIDES (text, background, images).

Question 2: I used several fonts in the same point size, but I noticed that some are more readable than others when projected. Why is that?

To explain this, you need to understand about ascenders and descenders, and about the x-height of a letter.

In Fig. 6.1, the ascender of the letter "d" ascends above the loop of the "d" and the descender of the letter "p" descends below the line. The letter "x" has no ascender or descender. Its size is representative of half the letters of the alphabet (a,c,e,m,n,o,r,s,u,v,w,x,z). It is called the x-height of a font. All things being equal, the greater the x-height of a font, the more legible the font is at a distance. For example, in Fig. 6.2, even though the two fonts have

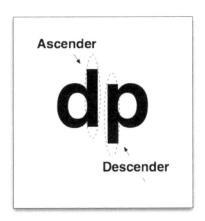

Fig. 6.1. Ascender of "d" and descender of "p" are surrounded by a dotted line.

Fig. 6.2. Two different fonts of the same size have different x-heights. The greater the x-height is for a given font size, the greater readability at a distance will be.

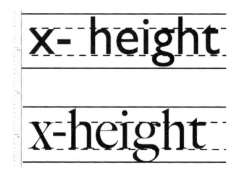

Fig. 6.3. Gil sans font on top and Times font below. The two fonts have identical x-height and font size, yet they do not have the same legibility at a distance.

the same point size, their x-heights differ. As a result, Font Gulim on the left will be more legible at a distance than font Gil sans on the right.

Besides x-height, the thickness of a letter stroke also influences legibility. In Fig. 6.3 the two fonts have the same x-height and the same font size; However, the Gil sans font on top will be more readable than the Times font below because its strokes are thicker. At a distance, the thin lines that make up the horizontal bar of the letter "e" will become invisible. The letter "e" will be mistaken for the letter "c", and the letter "x" will look like a backslash "\".

Question 3: People write that it is bad to write words in all caps because it is more difficult to read, but letters in all caps are bigger and should be easier to read. So why are they more difficult to read?

The eyes concentrate on the upper part of a word to decipher it. They rely more on the upper part of a letter (ascender) than the descenders. Maybe because there are more ascenders (b,i,j,d,f,h,k,l,t) than descenders (g,j,p,q,y), and because the frequency of the letters with descenders in a word is on average half that of letters with ascenders . The following example where the sentence is cut in two halves demonstrates this nicely (Fig. 6.4).

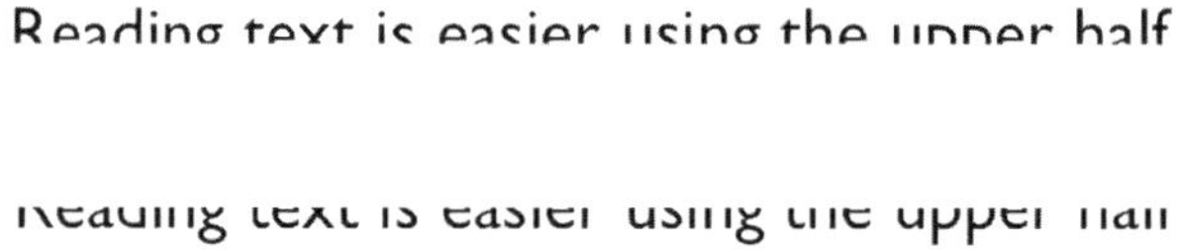

Reading text is easier using the upper half

Reading text is easier using the upper half

Fig. 6.4. The upper part of the text has been cut away from the bottom part. Notice how much easier it is to decipher the text using the upper part than using the lower part. Ascenders help decipher text quickly.

READING TEXT IN CAPITAL IS TOUGH

READING TEXT IN CAPITAL IS TOUGH

Fig. 6.5. Capital letters have no ascenders or descenders, making them difficult to decipher rapidly.

In all capitals mode, each letter of the alphabet has the same x-height. There are no more ascenders and descenders. Their disappearance prevents rapid identification of a word by looking at its silhouette. Reading is slower, even if legibility may still be fine when all capital letters have thick letter strokes (Fig. 6.5).

Question 4: People also write that we should not capitalise the first letter of each word in a title. Why is that?

Do not capitalise The First Letter Of Each Word In Your Slide Title Thus Forcing The Brain To Stop At Each Word For An Artificial Reset. It does slow down reading even though legibility may still be fine.

Question 5: I read that we should not use serif fonts in PowerPoint presentations. What is a serif font and why shouldn't we use it?

Serifs, (Fig. 6.6) are the little underlines at the foot of some letters (for example, the r, i, and f letters in "serif"). They create a semi-continuous horizontal line that guides the eyes when reading small prints and they connect the letters within a word.

Serifs fonts are mostly found in printed material where font size is small (12 points or less). Serif fonts have letter strokes of unequal thickness. In a presentation, however, text has to be large and letter strokes uniformly thick for maximum legibility at a distance. Sans serif fonts like **Verdana**, **Arial** or **Helvetica** are preferable.

Fig. 6.6. The serif: the small horizontal line(s) at the foot of the letters. For example, serifs are in the letters f, h, i, k, l, m, n, p, q, r, and x in this font.

Question 6: I always seem to run out of space on my slide when I use large text. So what can I do to put more text and still remain legible?

There is no easy answer to that question. The best way is not to put more text but to put less text on the slides. Therefore, when text is necessary, it is worthwhile to spend time rewriting the text to make it more concise while retaining the clarity of meaning. The natural tendency of presenters is to find a technical solution to the text-won't-fit-in-that-space problem. To put more text into the same limited space, they cram the text using a number of tactics. Most of them are bad.

1) Do not decrease the spacing between the letters forming a word by using the condense effect in the PowerPoint format/font menu... to cram more text onto one line so that it fits without having to choose a smaller font size. Instead, use a font that is condensed by design and remains legible such as this **futura condensed font**, or this Abadi MT condensed font.

2) Do not decrease the space between lines. See what happens when you... decrease the space between the lines thus squashing the lines to save space using PowerPoint's line spacing settings (Fig. 6.7).
   Giving enough spacing between two lines is necessary because it is the blank area around words that allows the eye to focus on them.

3) Do not use the PowerPoint option to automatically reduce the font size to fit text into a given space. You want to be the one in control of text size and not let Microsoft PowerPoint decide for you. Microsoft PowerPoint

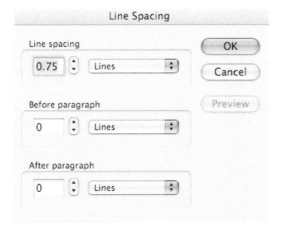

Fig. 6.7. PowerPoint [Format/Line spacing...] setting dialog box.

does not care if reducing the font size automatically reduces legibility at a distance. That is why, if not already done, the first thing to do when launching PowerPoint is to change the default settings. Microsoft Office differentiates title text from body text, and you can control the autofit separately for each in the *autocorrect options* of the *tools* menu. Unfortunately, these settings are set differently in the various versions of PowerPoint. The following Microsoft tip is given online for Powerpoint 2003: "About automatic text formatting".

 http://office.microsoft.com/en-us/powerpoint/HP030714011033.aspx

For PowerPoint 2007, check out the two URLs given here, under:

"Manage text in a placeholder by using AutoFit"
http://office.microsoft.com/en-us/powerpoint/HA101779661033.aspx
"Turn AutoFit text on or off"
http://office.microsoft.com/en-us/powerpoint/HA100475531033.aspx

4) Do not save space by removing prepositions and turning a clear phrase into an obscure compound noun. In the following example, after the two-letter preposition "*to*" was removed to gain space, the title became unclear and slow to read.

Transient model for kinetic analysis of electric stimulus-responsive hydrogels (unclear)

Transient model for kinetic analysis of hydrogels responsive to electric stimulus (clear)

Question 7: What is wrong with letting Microsoft PowerPoint do the word wrapping for me?

Word wrapping is very convenient when typing long paragraphs of small text. In presentations, text is large and word wrapping occurs frequently. Unfortunately, it is done without taking the meaning of the words into account.

The famous sentence used in examples on the role of punctuation will make the problem clear.

A woman without her man is nothing.

A woman without her man is nothing.

Meanings can change or become obscure when the sentence is broken at the wrong place.

When keywords in your title are split between two lines, legibility may be fine, but your title may become difficult and slow to read. Here is the title of the article published in *BMC Genomics*. 2007 Jul 3;8:210 by Yap DY, Smith DK, Zhang XW, and Hill J. The words in bold should be kept on the same line, and the long compound phrase in underlined italic should be clarified.

> Using biomarker signature patterns for an **mRNA molecular diagnostic** of *mouse embryonic stem cell differentiation state*

The same title is clarified with the two-letter preposition *in*. Word wrapping is still enabled but a more condensed font (Helvetica CY) is used to keep font size and length of line constant so that *diagnostic* would not be separated from its qualifying words *mRNA molecular*.

Using biomarker signature patterns for mRNA molecular diagnostic of stem cell differentiation state in mouse embryo

You can turn off word wrapping temporarily in PowerPoint 2003: "Wrap text in a shape or a text box".

 http://office.microsoft.com/en-us/powerpoint/HP051948231033.aspx

For PowerPoint 2007, "Position text in a shape or text box in PowerPoint 2007". http://office.microsoft.com/en-us/powerpoint/HA101326591033.aspx

In Apple Keynote, wrapping is automatic. You will have to watch when Keynote starts the word wrap and determine which strategy would work best to keep related words on the same line: rewrite the line to make the problem disappear, or manually press shift-return (shift-enter) before wrapping starts, or manually hyphenate the word and press shift-return before it wraps.

If you type return (enter) instead of shift-return, the software may automatically capitalise the first word on the new line, and that may not be what you wish.

Question 8: Are there any problems using unusual fonts, unusual styles or unusual alignments in presentation?

Unusual alignments create problems on slides particularly because text is large and fewer words fit on one line. Right alignment, and justify for example, dissociate text.

1) Do not right-align text. It is difficult to read because the eye cannot predict where the next word is going to start.

2) Do not introduce spaces of irregular size between words by aligning large text with justify instead of align left, thus creating "rivers" of white space between lines.

3) *Italics are not recommended.* They are more difficult to read than plain style (but some combination of font and background colours could change that — see the article "Readability of websites with various foreground/background colour combinations, font types and word" at the following website:

http://hubel.sfasu.edu/research/AHNCUR.html

4) *Do not use cursive writing. Cursive writing is easier to read for the French who study it even before block writing but readers of other nationalities find it more difficult to read than* print type.

5) Do not use exotic fonts. They may not be in the computer used for the presentation. If you write formulas prepared in another Unix text editor, your deltas and gammas may not convert to PowerPoint. You might want to turn your formula into a picture, or rewrite the formula using the "symbol" font. If you use exotic fonts, you must embed them into your PowerPoint file. If you do not, PowerPoint will substitute fonts with different width that may destroy your careful slide layout.

Follow this link http://pptfaq.com/FAQ00076.htm to find out how to embed fonts if you run windows, but be aware that the folks with a Mac will not see your fonts and that you cannot embed all fonts (only fonts with the file suffix *ttf* — True Type Font). Mac users can look here http://www.macosxhints.com/article.php?story=20051216145914724 to find out which embedded fonts are causing them so much trouble.

Question 9: In a nutshell, what should I remember in order to avoid legibility problems with text?

To summarise, best legibility of text at a distance is a matter of 1) its size: the larger, the better, 2) the x-height of the font: the taller, the better, 3) the thickness of the straight and curved strokes forming the letters: the more uniform, the better, and the thicker, the better, 4) the spacing between the letters within a word: enough to avoid muddling, but not so much as to induce dissociation, 5) the spacing between lines: the greater the x-height of the font, the thickness of the font, or the longer the line, the greater the interline space, and 6) the simplicity of its alignment: left-aligned, plain and bold is preferable to justified, right-aligned, centred, italic, underline, and all caps.

Question 10: What font size should I use for my titles to be legible at a distance?

This question is extremely difficult to answer. So much depends on what is outside of your immediate control. However, we could learn from people who design road signs. They set a minimum letter height of 10 cm. The fonts they use have to be very legible in various weather and sunlight conditions, so they use a sans serif font, with broad letter strokes and high x-height such as Helvetica or Univers. If you have the time, read the interesting article on the design of a new font found at the following URL

http://www.nytimes.com/2007/08/12/magazine/12fonts-t.html?pagewanted=1&ref=magazine

Driving speed is not an issue for a typical audience sitting in a meeting room. Yet, we could still consider 10 cm as a suitable font height for a highly legible projected title, in a large meeting room (from 15 to 20 metres deep). Translating this to the corresponding computer screen font size should be simple, but it isn't. It depends on the font x-height, your computer screen, the projection screen, and the viewer's eyes.

In 1988, as personal computer screens started to spread their green and white glares in a growing number of homes and offices, the human factor society recommended letter height for the highest legibility on video monitors. Naturally, the organ of sight is the eye, therefore optimal height had to relate to the height of the image formed on the retina. The recommended letter height corresponded to an arc of 20–22 minutes subtended on the retina (Fig. 6.8) — a minute is a sixtieth of a degree.

You can do the maths. The distance from the eye to the screen is the radius of the circle, and you want to know what is the length of an arc subtended by 20 minutes of arc, or 1/3 of a degree. Here is a graph that plots the relationship between the distance to the screen and the height of the letter corresponding to an angle of 22 minutes of arc (Fig. 6.9).

Now that we know the height of a letter for a given distance between the back of the meeting room and the screen, we still have to determine what is the corresponding font height on the computer screen (Fig. 6.10).

Here again, the maths will help us. A given screen size (expressed by its diagonal length) can easily be mapped onto a projection screen if both of them have the same aspect ratio (usually 4:3). The magnification factor is simply the ratio between the diagonal of the meeting room screen and that of the computer screen (Fig. 6.11).

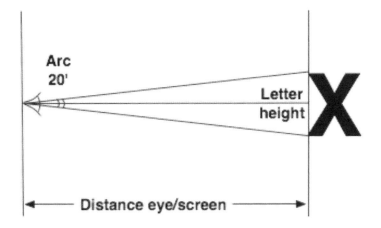

Fig. 6.8. Optimum angle for high legibility of image projected on retina.

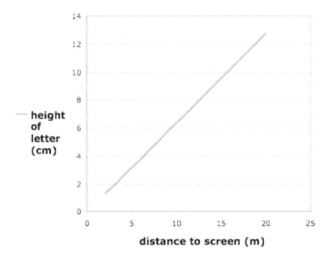

Fig. 6.9. Relationship between height of letter and distance from eye to projection screen for optimum legibility.

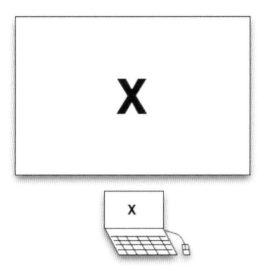

Fig. 6.10. Mapping of computer screen to projection screen.

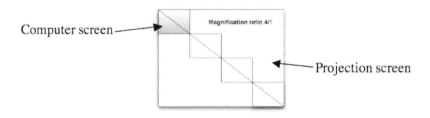

Fig. 6.11. Magnification ratio is between screen diagonals.

Distance from screen to back of meeting room (m)		2	3	4	5	6	7	8	9	10	15	20
		Helvetica font size for 15″ Mac MacBook Screen running PowerPoint (font size less than 12 or more than 70 is greyed out)										
	1	62	90	124	152	181	214	243	276	305	457	610
	2	31	45	62	76	90	107	121	138	152	229	305
	3	21	30	41	51	60	71	81	92	102	152	203
	4	15	23	31	38	45	54	61	69	76	114	152
	5	12	18	25	30	36	43	49	55	61	91	122
Magnification	6	10	15	21	25	30	36	40	46	51	76	102
ratio	7	9	13	18	22	26	31	35	39	44	65	87
between	8	8	11	15	19	23	27	30	35	38	57	76
projection	9	7	10	14	17	20	24	27	31	34	51	68
screen	10	6	9	12	15	18	21	24	28	30	46	61
and	11	6	8	11	14	16	19	22	25	28	42	55
computer screen	12	5	8	10	13	15	18	20	23	25	38	51

Fig. 6.12. Table to identify optimum Helvetica font size for 15″ MacBook used for projection on a screen of given size placed at a given distance from a viewer.

Therefore, with a known projected font size, and a known distance between screen and back of room, we can find out what is the height of the font size on the computer screen. For this simulation, we will take a 15″ computer screen.

The following table (Fig. 6.12) is valid for a Macbook running PowerPoint in default mode (720 by 540 pixels). Fonts that are too small or too large for a presentation are greyed out.

For example, take a conference room with a computer screen/projection screen magnification ratio of 10, and a distance from screen to the back of the room of 10 metres. In this case, Helvetica font size 30 ensures optimum legibility.

Note that this measure is valid for PowerPoint only. You would only get the same results with Apple Keynote if you set the screen size to the custom size of 720 by 540, which happens to be the default size in PowerPoint.

To derive the table, it was necessary to determine the relationship between the size (in points) of a particular font and its height (in cm) on the screen of the computer. Here is how to do it on your computer for a font type of your choice. The example given here is for the 15 inch Macbook and the font is Helvetica.

1) Open PowerPoint.

2) Create a blank slide with a white background.

3) Type the letter "n" in sans serif font Helvetica (like "x", it has no ascender or descender).

4) Set the size of the font until top and bottom of the letter fit with the top and bottom of your slide.

5) Go into slide show mode and view the slide.

6) Measure the height of your image (it should be the same as the height of your computer screen).

7) Divide the font size by the height of your image in cm and you get the number of points per centimetre.

Table 6.12 gives the font size for headings, but what should be the font size for the text under the title bar of a slide? A font size between 20 and 30 is fine as long as the font is sans serif with a high x-height. Do not go down to font sizes below 20, and if you absolutely have to, use bold to enhance legibility, expand the spacing by selecting the expanded option in the format/font menu, and always increase the spacing between lines. Letters that are further apart (but not too much) are easier to identify at a distance since they do not merge so readily with their neighbours.

Question 11: How do I know whether the projected image will be legible before the event when I do not even know what kind of projector will be there, and how deep the conference room is?

This is indeed a very good question. After all, isn't your computer screen there to help you identify potential colour and readability problems before the conference? Actually, no. Indeed, when you prepared your slides on your computer, everything on the high-resolution screen was legible. The green fluorescent protein marker glowed brightly on your high colour fidelity 15-inch glossy computer screen. Even point 8 fonts were visible on your 1440 by 960-pixel screen (one million and three hundred eighty two thousand and four hundred pixels). The audience, however, faces a matte screen displaying only a third to a half of the pixels on your PC screen. The audience is the one with the legibility problem, not you. Some colours on the big screen are washed out, others scintillate and vibrate uncomfortably, and others do not even appear. Your computer screen gave you a false sense of confidence. You trusted it while preparing your perfect presentation, but the trust was misplaced. The betrayal is collective. Many factors contributed to it: the projector, the distance that separates the audience from the big screen, the ambient light in the meeting room, and dare I say, you also.

Let us first tackle the problems linked to projector and audience. A typical projector has less resolution (SVGA: 800 by 600 pixels, and XGA: 1024 by 768 pixels) than your computer screen (1440 by 960 pixels). However, the advantage in resolution that your computer screen has over the projector disappears as soon as you start looking at both screens from a greater distance. This is why people who buy SVGA projectors claim that there is no noticeable difference with an XGA projector… as long as people step back 10 metres from the screen. Closer to the screen, you would actually see a difference. People sitting in the front row have less legibility problems than the people sitting at the back.

Knowing this, you can simulate at home what happens when your audience sits far away from the screen. Just sit away from your computer screen! Stretch your arm in front of you, palm horizontally stretched (Fig. 6.13). If you touch the screen when you do that, it means you are sitting too close! Now, staying in that position, slide your chair back until your palm hides the whole computer screen. You are now at the right distance away from your screen to simulate what the audience sees from the back of the room.

Notice that the screen is not as bright. Adjacent pixels merge and thin lines blur. Pastel or fluorescent colours fade or disappear. And text too small

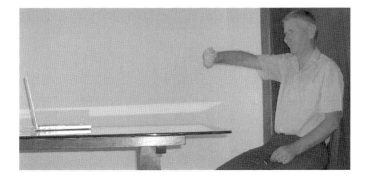

Fig. 6.13. Method to investigate legibility at a distance.

becomes illegible. You will catch most (not all) legibility problems early in front of your computer at home, instead of in the presentation room at the last minute.

There are additional readability problems due to the process of matching a high-resolution image to a lower resolution image. The projector scales or compresses the image. The new projectors do this quite well. The older models lose some detail in the process. Keystone correction adds further detail distortion. Therefore, to be safe, you may want to move that chair of yours back a little more.

Question 12: When I start presenting, the meeting room lights are usually turned down. Is that good?

My purpose is to turn you into a great presenter. For that you need to remain visible and keep eye contact with the audience. You also need to stay away from lighting conditions that will give your audience an opportunity to doze off!

Question 13: If I keep the meeting room lights on, are my slides still readable and what background colour is more appropriate?

Keeping the lights on does create legibility problems. You need to be aware of them to avoid them and design your slides accordingly. The good news is that if your presentation is legible in daylight, you are ready to face the impossible. Daylight washes out the projected image. You can simulate the effect of daylight at home on your computer, and here is how.

1) Choose a slide that you suspect might have a problem in high light level.

2) Save as (choose the name "light-test" if you want), and in the dialog box, change the default format from PowerPoint to the PNG format (Portable Network Graphic), and click the option button to reveal another dialog box where you select the buttons as seen in Fig. 6.14. In this case, since the projector is a SVGA projector, I selected the size 800 by 600. You would change these to 1024 by 768 for XGA.

3) Open the file in any graphic editing package, Adobe Photoshop or others (I use the GraphicConverter software on the Mac).

4) Select the menu option that allows you to change the brightness levels… and experiment.

Fig. 6.14. PowerPoint [Save as.../options] setting dialog box.

This is what you will discover while experimenting the effect of an increase in brightness in the room:

Pastel and light grey colours just fade away as you increase the brightness. The black background washes out to become greyish. Plain text thins out as the brighter white pixels gang up against the lonely black pixels. But do not stop there. Now that you have your text slide, log on to the colour-blind test website,

 http://www.vischeck.com/vischeck/vischeckImage.php and for PC users http://www.fujitsu.com/global/accessibility/assistance/cd/ and submit your image to discover what people who are colour-blind really see. You are probably in for another surprise.

For the use of colours, Microsoft has an excellent webpage available at the following address:
http://msdn.microsoft.com/en-us/library/aa511283.aspx

People have debated over the choice of background colour and the interaction between text colour and background colour. See
http://hubel.sfasu.edu/research/AHNCUR.html
as well as the discussion on background colour on Edward Tufte's website.

116 Slide Design

http://www.edwardtufte.com/bboard/ q-and-a-fetch-msg?msg_id=000082

I recommend the use of a white background and black text.

Question 14: How do I avoid the problems caused by the use of a white slide background?

Your pale colours will wash out. Colours over a white background need to stand out. For that, they need to be darker, not lighter. Therefore, use saturated colours. Saturation is defined as the degree of difference of your colour from white. You can always increase the level of saturation of a colour by bringing that level closer to pure black in your colour wheel (Fig. 6.15).

The best way to avoid colour problems is not to avoid colours, but to understand colours and to use them properly. For example, if you have to write text inside a saturated dark colour box, which colour text will you choose (remember the definition of saturation)? The answer is a colour with low saturation or white. Now the opposite is true also, if you have to place coloured text on a light colour background, use text with saturated colour or black (Fig. 6.16).

Remember that, from a distance, things tend to merge unless they are far apart or quite different in colour. Therefore, do not colour text or objects using

Fig. 6.15. Colour wheel. When right slider is pulled down, colour saturation increases.

Fig. 6.16. Evidence of legibility problems associated with colour saturation.

two adjacent colours in close proximity on the colour wheel; your audience may not be able to distinguish the difference between them.

Another problem you may face with a white background is the actual brightness of the projected image. A projector that is too bright may give some people a headache if your presentation is long. To reduce the contrast between the screen brightness and the room brightness, make sure the technician does not dim the lights prior to your presentation, and give visual relief by using the "B" key from time to time. You could also adjust the brightness controls on your projector and lower the brightness to reduce visual strain.

One last problem occurs when your visual displays fluorescent protein markers. These are usually only visible in low room light and over a black background. When you have such a visual in a slide, lower the light level in the room (yourself or ask the technician), but only for that slide.

Question 15: Pasting an image into Apple Keynote does not give the same results as pasting the same image into Microsoft PowerPoint. Why is that?

Keynote and PowerPoint have different behaviours. Keynote is pixel-based, while PowerPoint is size-based. In Keynote, when you open a new document, in the opening dialogue, (Fig. 6.17) a sub-menu next to slide size proposes five sizes (custom sizes are also possible):

600×800: SVGA projectors
1024×768: XGA projectors
1280×720: WXGA high definition home cinema projection
1680×1050: Wide Screen Wallpaper
1920×1080: WUXGA True high definition cinema projection

It is a good idea to check the general preferences under the Keynote menu. By default, the editing preference should indicate, "Reduce placed images to fit on slides". Keynote downsamples the image (permanently) to fit it on the slide. Keynote takes into account the slide (pixel) size you have chosen and it fits your image into the slide as long as the image pixel size equals or exceeds that of the slide. For example, if you have selected 1024×768 for your slide size, and you have four images: image A (640×480), image B (800×600), image C (1024×768), and image D (2048×1536). When dragged into your empty slide, image A and B are smaller than your slide.

118   Slide Design

Fig. 6.17. Keynote [Inspector/document inspector/document] setting dialog box.

Image C fits exactly into your slide and image D is scaled down to fit perfectly into your slide because its aspect ratio (3:4) is the same as your slide aspect ratio. Keynote scales the image so that it always fills the screen, but hides the rest (Fig. 6.18). The hidden part is still there, you only have to grab the image and move it sideways, or up and down. If the aspect ratio is different from 3:4, with more pixels than needed (more than 1024, or more than 768), Keynote scales the image so that one side of the image fills the width or height of the screen while the other side fills more than the screen. The image would not be forced into the screen size through distortion. For example, for an image with a 7:5 ratio (as opposed as the regular 4:3), 5 would be brought down to 3, and 7 down to 4.2, the 0.2 would be hidden. If the aspect ratio is reversed (vertical image), for example 5:8 (as opposed to 4:3), this time 5 would be scaled down to 4, because if 8 was scaled down to 3, 5 would proportionally become smaller than 4 and the image would not completely fill the screen.

PowerPoint behaves differently. It gives the slide size in inches. The default size is 10″ by 7.5″ (notice the standard aspect ratio of 4:3). You can verify for yourself by setting the ruler units to inches in the preference menu and then turning on rulers in the view menu.

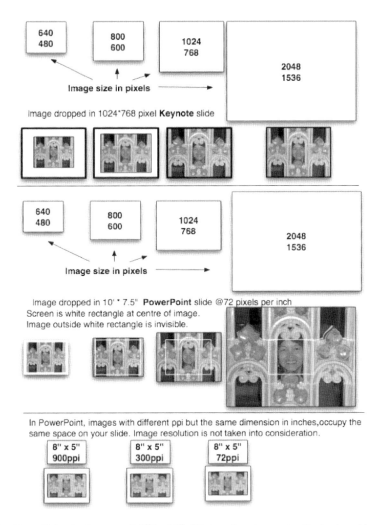

Fig. 6.18. PowerPoint fits image on 7.5″ by 10″ slide based on image size (inches), and ignore image resolution (ppi); Keynote fits image on slide based on the number of pixels in the image. Pixel-based, Keynote does not ignore image resolution.

You can change that default 10″ by 7.5″ size in the File/Page Setup dialog box (Fig. 6.19).

Unless you specify your image size in cm or inches in a graphic software application (say in Photoshop), the pixel to inch conversion is done by dividing the number of pixels of your image by the standard 72 pixels per inch (ppi). In PowerPoint, image A would be 8.88″ by 6.66″, image B would be 11.11″ by 8.33″, image C would be 14.22″ by 10.66″, and image D would be 28.44″ by 21.22′.

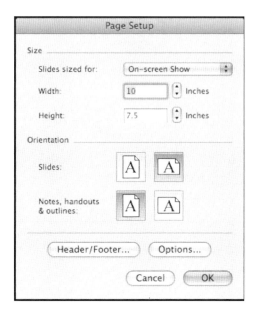

Fig. 6.19. PowerPoint [PowerPoint/File/Page setup...] setting dialog box.

As expected, image A is smaller than the 10″ by 7.5″, and is therefore smaller than the slide. Image B, which respect the 4:3 ratio, is larger than 10″ by 7.5″, will PowerPoint scale it to fit the screen? No, unlike Keynote, it puts the centre of the image at the centre of the screen and hides whatever drops out of 10″ by 7.5″. As a result, since the overall image is larger (see Fig. 6.18). PowerPoint in effect cuts a fixed window inside the centre of the image. The larger the image, the more detail is in that centre window, giving the illusion that we are zooming in when in fact, we are not.

Now that you are aware of what both programs do, you will not be surprised by what happens when you bring in your visual into a slide. The larger the image size and the greater the number of pixels (at least 1024 pixels for the image width), the more your image is legible. If your image is small but has a high number of pixels per inch (300 ppi for example), you can increase its size through a reduction in its resolution, but do not reduce below 72 ppi (Fig. 6.20).

Apple Keynote makes it easy to bring in large photos and fit them in your slides. Microsoft PowerPoint, however, makes it difficult if the size of the photo is larger than 7.5″ by 10″. To alleviate the difficulty, place your image anywhere inside your PowerPoint slide. After selecting

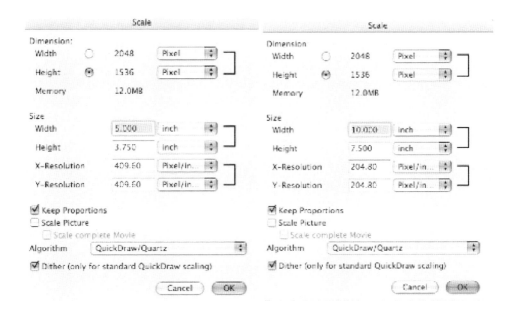

Fig. 6.20. GraphicConverter [Picture/size/scale...] setting dialog box.

"Insert/picture.../from file", once the picture appears on your slide, double click on the photo (Mac) or right click on it (PC) to bring the format picture dialogue box. Select the position tab and put zero into the horizontal and vertical position windows. Then select the size tab, and type whatever size you want the picture to be, but make sure that the aspect ratio is locked so as not to distort the picture.

Question 16: My pictures have many colours or grey levels, my diagrams are quite complex, and my graphs have many curves and much detail in text legends. When I stand back and look at my screen, much of these details become unreadable. What can I do to keep things legible?

Bringing a picture in the presentation software is one thing. Making sure that it is legible is another. Scientific pictures are often very detailed. Material science relies on colour scales to express a range of temperatures or pressures. In every field, rare are the graphs with only one curve. The multiple curves that usually crowd graphics need to be individually identified with small text. Alas, small text is not easily seen once projected on a large screen. Graphs exported from a scientific visualisation package or screen dumps have tiny greyish legends written in serif fonts. Once projected, the letters look thin and anaemic. The one-pixel lines of your graphics look anaemic too. The numbers on the x- and y-axis are in font size 8 or 9. The dark text is barely noticeable on the saturated colours that come standard with the scientific software. The white scale on your Transmission Electron Micrograph disappears over the brighter areas. You suspect that the fluorescent green or red will simply be invisible in daylight. In other words, everything that looked fine on paper or on your high resolution monitor, now appears totally unsuitable for screen projection. What can you do?

Someone has to be the bearer of bad news. You need to redo your visuals to rid yourself of all legibility issues. This is time consuming work. However, the stakes of a scientific presentation are high and certainly worth the effort. Often times, the scientific visualisation package lets you set drawing parameters to improve legibility such as font type, font size, line thickness, line colour, or placement of legend. Redo your visuals and decrease the number of curves keeping only the key curves that make your point. The other curves can be shown later, if necessary, on a supplementary slide placed after your conclusion.

If you use Microsoft Excel for your charts, did you know that you could change practically everything through the various dialogue boxes, font size, font type, colours, background, etc? You can even add text, and change the way the curves are labelled by bringing the meaning of each curve close to the curve, legibly, instead of in a detached curve legend with text so small, it cannot be read once projected. Remove the default grey background underneath your black curves because it decreases legibility!

Redo your curves using the PowerPoint graphic tools (or another graphic software) and import the final graphic into PowerPoint as an image. Sit away from your screen to check the legibility of your work. Remember that it has to be seen by the scientists sitting at the back of the meeting room, and in a bright light.

Increase the contrast and brightness of your pictures (within PowerPoint): the higher the contrast, the higher the legibility, research has shown it. In PowerPoint, double click on the image (Mac) or right click the mouse (PC). The contrast and brightness controls are under the tab "picture". In Keynote, it is the "adjust" button on the top bar of the application. Keynote lets you adjust more than brightness and contrast (Fig. 6.21).

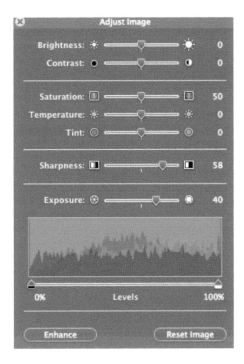

Fig. 6.21. Keynote top bar Adjust image button.

Question 17: Is it important to align objects on my slides?

Absolutely. Symmetry pleases the eye. You could say that the alignment of objects is like the cherry on the cake. It is the final visual touch of class enhancing the pleasure of discovering your slide contents. There will always be a fraction of the audience distracted by your lack of care when placing text or visual objects on a slide. These scientists usually have neat slides. They like everything to be neat. When in a restaurant, they usually correct the placement of the knife and fork if not perfectly parallel. They move things around until they look orderly. They take time and get great satisfaction out of aligning things. Anything out of alignment on your slides will irritate them. They will notice that the title on one slide is lower than the title on the previous slide, for example. They will think you are an untidy, carefree sort of person.

PowerPoint and Keynote both provide many tools to align. If you type *Align* in the search box of Keynote Help, under the main help menu, you will see the following list (Fig. 6.22).

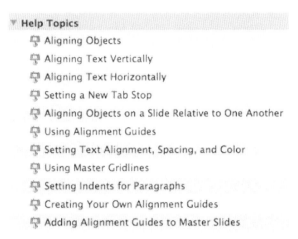

Fig. 6.22. Topics retrieved by Keynote Help on the search topic "Align".

In passing, notice how Apple's human interface facilitates reading the items on a list by alternating the background colour under each list item. The colour separates each item from the surrounding items. Such a technique is useful, for example on your conclusion slide, when you list out the main items covered during your talk.

Whatever the software, alignment techniques include gridlines, guides, the layout of the elements on the Master slide template, and alignment commands provided by the software, such as *snap to grid*, *distribute objects* horizontally, vertically or *align objects* left, right, centre, top, middle, or bottom.

Alignment is not only eye candy. It guides reading order. People will usually read from left to right and from top to bottom if all text objects are of the same size. They will read an item in bold before they will read an item in plain text. They will read horizontal text before vertical text. All this is predictable. The alignment, font style, and direction of your text will more or less fix the reading order. Do not let people's eyes wander through your slide. Decide on the optimal path the eyes should follow to extract the information in the right order.

Design for Audience Attention

The audience came to listen to your talk. Therefore, you have their interest. But do you have their attention? It depends as much on you as on your slides.

I would even argue that it is impossible to dissociate presenter and content. In this slide design section, the two are related, as they should be.

Critical Things to Know About Attention and How It Influences Slide Design

Attention is actively split across multiple concurrent channels

Attention is active. It uses different channels concurrently. Attention is about listening and looking (as opposed to hearing and seeing). Both senses collaborate by locating and focusing on the same source of information in order to check, confirm, interpret, and act on the information. For example, when you smell cigarette smoke in a smoking-free zone, you scan the area in search of someone smoking. Or when people tell you "I give you my undivided attention", you know that what you say has a greater probability of being received without loss because sound and vision are both in alignment.

How is attention strained? When it is divided. If the two senses are pulled in different directions, even though each sense has its own independent capacity to pay attention, inefficiencies occur. Car accidents caused by people driving while on the phone are a dramatic illustration of the consequences of a divided attention. Speech unrelated to what the driver does creates conflicts of attention that are difficult for the driver to resolve. In presentations, speech unrelated to what the slide contains creates conflicts of attention that are difficult for the audience to resolve. As is often the case, eyes take precedence over ears, and the audience mostly looks at the slide and reads, rather than listens. Therefore, when you need to depart from what the slide supports, it is better to blank the screen using the "B" key or by fading to black.

One may think that when there is total redundancy between audio and visual channels, attention is not strained. For example, one would think there is maximum synergy when the oral content is a straightforward reading of the written content. However, because the eyes can read a text faster than that text can be spoken, the audience is frustrated and turns down (or off) the oral channel, paying episodic visual attention only when a new slide appears and welcoming distractions at other times to avoid boredom. This redundancy is also possible within the same visual channel: text can describe what anyone in the audience recognises without text. For example, a photo of a zebra fish is shown, and the text says "zebra fish". For some audiences engaged in life

126 Slide Design

Fig. 6.23. Is this an E or is this an F?

science work, this is akin to showing the picture of a cat and adding the word "cat" next to it. The audience expects to be shown something new.

Attention autonomously moves from global to local

Attention starts wide, at the global level. In Fig. 6.23, the eyes will see the letter E before they see the letter F.

Attention finds a match with what is already known to ease the interpretation and the extraction of differentiating details. Scientific visuals are compared with a class of known objects that come with their pre-wired sets of expectations, and interpretation rules. For example, as soon as the brain recognises a pie chart, it actively explores the visual by following a preset succession of activities. It looks first at the title of the chart, then at the most important section of the pie starting clockwise from the top. A bar graph would not be "looked at" the same way as a pie chart or a scatter graph.

Attention is autonomous but it is also docile. We will see later that the presenter can always override the preset pattern of explorations by setting attention traps in specific areas of the chart (using colours or other attention-enhancing devices) or by guiding the eyes using words or a laser pointer.

Attention is triggered by change

Even though our brain can channel our attention by forcing our eyes and ears to focus on what is interesting or what requires more brain cycles to process, our eyes and ears can signal the brain to shift attention to new events. Attention is captured by what is new (a change from the old), important (a change through emphasis), what changes (colour, shape, sound) or announces a change (blank screen — "B" key, transition word, silence, sound), what appears and what disappears, what moves, stops moving, or changes direction. New information occurring outside our centre of vision, particularly through motion, is sufficient to shift the focus of our attention. Furthermore, our eyes are fast to hunt for new targets and slow to return to past-explored targets, a phenomenon known as *inhibition of return*.

To be an object of attention, the target must first stand out from its surroundings. Leaving enough space between objects on a slide helps identification. So does using different colours to formalise the separation between items. This last technique is particularly effective to make the points on a conclusion slide stand out: use an alternate font colour (dark blue and black, for example) for each conclusion point.

Attention is triggered by questions and pending expectations

A question creates a vacuum that only an answer can fill. It sets the expectation of an answer. That expectation does not die; it lingers in the mind until it is fulfilled. The audience is kept attentive hoping to hear the answer, but it is also encouraged to think and explore various plausible answers (if given time to think or a range of possibilities to consider). A question does not always require a question mark. Strong statements are questions in and of themselves, particularly when they question our existing knowledge: "*The world is flat*" makes an intriguing title for the book of Thomas L. Friedman because the statement immediately questions the reader.

Expectations have been presented in Chapters 3 and 4. They are raised by the title of your talk and by the scientific process itself (no claim without proof). The following expectations are present at all times among people in the audience:

1) What is the contribution?

2) Is there a general need for such a contribution?

3) Do I, the attendee to this talk, have a specific need for this contribution? Would it solve a problem I have?

4) Why has this method (also tool/data) been chosen to tackle this problem? Is this the best method to do this?

5) What are the limitations of this work (assumptions, method, scope)?

From a purely scientific perspective, people expect their attention to be rewarded by the discovery of a new area of research, or the discovery of new or improved tools worth adopting to solve daily problems. Maybe the reward is a simple confirmation that one is on the right track, or a better understanding of something complex.

Listen to the "Three audience irritants" podcast on the DVD

Design Techniques to Capture and Sustain Attention

Emphasise

Information placed in **a slide heading is naturally emphasised.** The eyes see it first. It is written in clear sans serif text larger than the text on the rest of the slide. It is important to strategically use headings for featuring what is new and important in your talk. The heading "experiment results" is neither new nor important. Headings that are just pointers such as *outline*, *motivation*, *results*, or *methodology* have no place in your slides. They waste valuable space that should be used to deliver the key point, claim, or message of your slide. The heading can be a phrase without a verb, or a full sentence with a verb. My preference is a full sentence because there is less ambiguity and it is more active. For example, the sentence "Gene A enables B phenomenon" is more dynamic and clear than the phrase "Gene A-enabled B phenomenon. The full sentence should be as concise as possible. The sentence "Phenomenon B is enabled by the presence of Gene A" is too long. The words "the presence of" are unnecessary, as is the passive voice, a well-known sentence lengthener.

The text message given in a slide heading has to be identified in its supporting visual. Unfortunately, that message is hard to identify clearly inside a visual where many curves are stacked, each with its own label, colour or pattern. Therefore, if a curve carries important or new information, **point to it** and explain. You could also use any of the following alternatives:

1) Circle or frame the important area (use a line thickness greater than 2 pixels)

2) Point while explaining within a callout box

3) Mask the unimportant areas with a shape whose transparency is lower than 100%

4) Change slide and show a close-up of the important area

We already discussed emphasis in text, using **style, colour, and font size or font types**. Be conservative. Limit yourself to two or three, and once you have decided when to use this font, that colour, or that style, apply the same principle over the whole presentation to preserve consistency. If you choose dark red to alert, do not move to orange on the next slide for the same function. If you use dark red many times on a slide, it is like crying wolf too often: the audience no longer considers that colour important.

Since the important information is not easily found in a cluttered slide, create space by discarding what is less important. Surrounded with space, the remaining information stands out. If your background is white, you have one more tool to bring emphasis at very low cost: shadows. They bring out objects and text from the background. PowerPoint has two shadow tools in its graphic toolbar: one with settings (see Fig. 6.24 — the shadowed square at the bottom of the graphic toolbar) and another called "simple shadows".

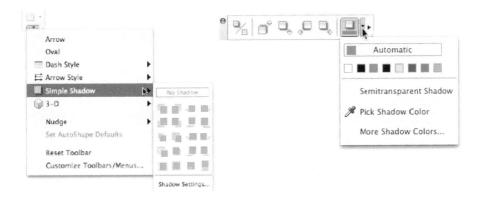

Fig. 6.24. PowerPoint [View/Toolbar/drawing/shadow/shadow settings] dialog box.

Fig. 6.25. Keynote [Inspector/graphic Inspector] shadow setting dialog box.

Keynote makes adding shadows very easy. Use the inspector window under the graphic tab (Fig. 6.25).

However, avoid using 3D to display a two-dimensional X/Y relationship simply because it is easy to do. Let 2D be 2D.

Besides the static highlighting techniques considered this far, emphasis is also given through motion (PowerPoint calls it "animation", and Keynote "builds"). Motion always attracts attention. But before we go any further, let us make clear that each movement on a slide or between slides should be justified and not "for the fun of it". In PowerPoint and Keynote, effects allow objects to grow or shrink in size, zoom in and out, fly in and out, drop in and out. These effects should only be used sparingly if at all in scientific presentations. The "appear" effect, on the other hand, is useful. It enables progressive information build up, and in particular, the overlay of an arrow pointing to the important data. Why use an arrow when you could use a laser pointer, you may ask? With a laser pointer, legibility is not guaranteed on a white background. The pointer's battery may be low or the screen too bright. It is safer to point with a screen object (arrow, callout-box) within PowerPoint or Keynote. The additional information or visual pointer (an arrow for example) appears after a mouse click or after a user-defined delay. Here is how to create a delayed build in Keynote and PowerPoint.

Keynote. On your slide, select the object you want to display separately from the rest (for example, an arrow). Select inspector in the top bar (the white I in a blue circle). In the inspector window, select the build inspector, a yellow diamond. The *Build in* button is already selected. Choose the *appear* effect in the Effect menu. Click the button *more options.* A *build order* drawer opens and your object is highlighted. At the bottom of the drawer under Start Build, select *Automatically after transition* or *Automatically after build #* … if there is more than one build. Finally, enter the number of seconds next to delay (Fig. 6.26).

PowerPoint. In PowerPoint: *Slide show > custom animation >* click on the object to bring in a layer under the text *select to animate*. Alternatively, you can click on that object as in Keynote, while still on the slide. It will appear highlighted automatically for you in the select to animate scrolling window. Click on the button *add effect.* In the new window that pops up, click *appear* under the entrance button — then click on the *effect options* button in the new window that just appeared, click on *timing*, and then under the *start menu*, choose *after previous* — and enter the number of seconds next to *delay* (Fig. 6.27).

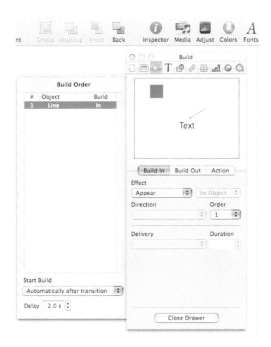

Fig. 6.26. Keynote [Inspector/Build Inspector/more options] effect setting dialog box.

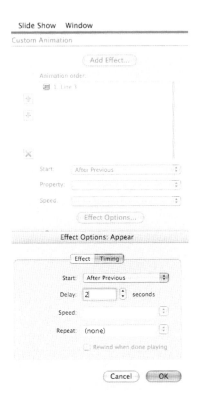

Fig. 6.27. PowerPoint [Slide Show/custom animation/Add effect] setting dialog box.

Animate images

Slide transitions have little purpose apart from three general instances, and one special instance:

1) For fade to black and fade from black effects to replace the "B" key effect (Fig. 6.28)

2) To transition from the visual outline (map) slide into the various outlined sections of the talk (for example using the cube transition if your outline has four parts)

3) To mimic the effect of layers by duplicating a slide multiple times and adding, subtracting, or modifying one or several elements on each of these slides

4) To use slide transition to reinforce the point made. For example, at one place in your talk, you may say, *this is the way things have been done so far,*

Fig. 6.28. PowerPoint [Slide Show/slide transition...] setting dialog box.

but our technology allows us to start a new chapter, to turn the page on the past. At this time, a one second *page-flip* transition in Keynote would definitely reinforce the message. In PowerPoint, you might say instead *it is time to uncover what the future has in store*, and use the *uncover left* transition with a long duration.

Video or image animations, on the other hand, bring a moving (pun intended) and well-needed support to scientific presentations, in particular when sequential processes are explained, when motion makes a static image come alive, and when one short video replaces one thousand graphics. Audiences love moving images. To be able to use audio/video media is a key advantage of a dynamic oral presentation over a static paper. Not using animations when they could shorten your presentation or convince the audience far better than a static visual is so short sighted. Yes, animations take time. Yes, a technical video is a challenge to do well and to keep brief. Yes, an animated gif file requires the proper tools to do well. Yes, you need to understand how PowerPoint or Keynote handles animations. Yes, animations and movies have a high potential for glitches given the great diversity of animation or video encoding formats. BUT YES ALSO, being clear and convincing is worth the extra effort. The stakes of an oral presentation are high. These extra hours spent on clarifying via animations are determining your future. Hasty short cuts do just that; they cut short a promising future.

PowerPoint (the PC version with motion paths) and Keynote (with actions) have built in tools for animations. Take time to learn how to animate your graphs and images, but do not limit yourself to these standard applications.

Become familiar with other animation software. PhotoToVideo on the Mac, for example, is like the poor man's video camera. With it, a static image turns into a dynamic one through all the standard camera and video editing techniques of the cinematographer: pan, zoom, rotation, titling, fades, transitions, audio. You basically guide the viewer into your image and highlight places of interest.

Animations are brilliant in the following cases:

1) To help the audience identify the start and end of a complex path (process), and to follow that path (process)

2) To show objects in motion (of atoms, molecules, vehicles, electrons, proteins, etc) or objects dynamically interacting with other objects

3) To emphasise the effect of one varying quantity (time, temperature, pressure, forces such as gravity) over another

4) To perfectly synchronise audio (speaker explanations) and video (slide content explained), helping the audience not to get lost in a complex diagram

5) To shorten a long video that is taken with a fixed camera (as in an electronic microscope)

Characteristics of a good animation

Choose your framing scheme. It could be a fixed frame inside which the objects evolve for the duration of the animation. The audience is at all times aware of all objects inside the frame, but focuses on what changes inside the frame (what enters and leaves the frame, as well as what moves within the frame). If you explore several frames, make clear that you are leaving one frame and entering another (as when entering a new room). A fixed frame is best used when there is no "star" object, and the frame dimensions represent a unit of some kind (a cell for example). Alternatively, the frame could be mobile because it is keeping a "star" object in its centre. What is observed in such frames is the interaction of the "star" object with its surroundings. You could blend several framing schemes to enhance interest.

A scientific animation is not a movie where the audience feels like going for a popcorn refill. It should be short, from a few seconds to half the time you spend on a slide. If long, the presenter should control the start and stop functions, and only show parts of the animation (if using Quicktime, create chapters).

An animation should only play once, at the request of the presenter (mouse click). Avoid automatic, looping animations that are terribly distracting. If you want to play an animation several times, ask the audience to focus on a different aspect each time it is played.

Finally, text in animations should be legible.

Design for Presenter Flexibility

File://localhost/Users/Vladimir/Desktop/VladPrezo.ppt

An hour before the board meeting, Vladimir's manager comes by Vladimir's office while he is rehearsing his presentation.

"I see you took good notice of what they taught you in your presentation skills course at the Maui Hotel. Good job. I hope you won't let me down this time. No pressure intended, of course."

"Of course", Vladimir replies, grinning, "but don't worry, I'll be fine. I'm a little nervous that's all. Want to look at my slides again?"

"Why? Did you change anything since I saw them yesterday?"

"No, I just added a few extras links to some documents outside the presentation, just in case they ask me technical questions; you never know."

"You never know", echoed the manager. "I'm glad you are taking this seriously. Professor Takayushi is a new board member. Just arrived from Tokyo last night. Impressive brain. Better be prepared. By the way, I came to tell you that we can finally load our presentations on the computer in the boardroom. The techies fixed the network glitch."

"About time", said Vladimir, with a sigh of relief. "We are supposed to start in fifteen minutes."

When Vladimir arrives in the boardroom, there are already three other presenters waiting for their turn to upload their presentation. Vladimir waits impatiently as they load and open their presentations to rapidly go through their slides. One even discovers he has taken the wrong file and rushes out of the room. One is taking his sweet time and looks as though he is rehearsing again. Everyone complains and he is forced to leave the lectern. When Vladimir's turn comes, there are still two people behind him. He downloads his presentation to the computer in record time, launches PowerPoint, and to his great satisfaction, the first slide appears exactly as it did on his computer. He quits and sits at the back of the room next to his manager. The board members arrive while the last presenter is still downloading his slides. After a short introduction by the director, the meeting begins.

When Vladimir's turn to present arrives, he gets up, moves to the lectern, gets to the first slide of his talk, and starts. After a brief introduction, he moves down his slides effortlessly, having well rehearsed. His manager sits relaxed at the back of the room. Suddenly, Professor Takayushi interrupts and asks, "Could you justify that?" Vladimir smiles. He had anticipated someone might ask the question and he had linked the slide to an external PowerPoint presentation.

"I am glad you are asking that, Professor. I have here a slide…" And as he pronounces the words after clicking on the transparent button hyperlinked to the file, he turns to the screen and freezes as he reads the laconic message: "Cannot open the specified file". He clicks one more time on the button to try his luck but the screen stubbornly displays the same message. He flinches as he remembers having only downloaded the main file, not the linked files. Without wasting time, he turns to the computer, quits PowerPoint, takes his USB drive from his shirt pocket, struggles to plug it next to the wireless USB remote control (he made a mental note to buy a thinner thumb drive next time), downloads the file, launches PowerPoint, returns to the slide (all this in record time) and clicks again on the transparent button. To his great consternation, the same message appears. At this time, Suzie, the techie, snaps into action, rushes to the lectern (while Professor Takayushi turns toward the Director and murmurs something that makes the Director laugh), she removes the old link to the external file which was still pointing to Vladimir's computer path names, sets the new path name, and goes into slideshow mode. It all took less than three minutes. Vladimir, blushing, apologises and clicks one more time on the transparent button. This time, the finger cursor changes shape but nothing happens. Vladimir is now in total panic. He looks away from the screen toward the Professor now engaged in a lengthy conversation with Dr. Linda Sinclair, another board member. As to his manager, he is shaking his head in denial and sheer frustration. In a trembling voice, Vladimir says, "I am sorry about this professor, maybe I can answer your question later". But just as he finishes speaking, the linked slide miraculously appears. It had taken ten seconds to appear, but ten seconds too long it appeared because the Director now addresses Vladimir coldly.

"Thank you Dr. Toldoff, but we have a full day of presentations ahead of us and we are already running late. Why don't you move to your conclusion slide?"

When Vladimir returns to his seat, his manager looks tense. In a low voice, he angrily barks "Why didn't you save your file as a PowerPoint Package? It saves all of the linked files in one folder!"

Sheepishly, Vladimir says...

"PowerPoint class?"

In many presentations, the arrangement of slides is as inflexible as the 35 mm slides arranged in a Kodak slide tray. There is only one way through the slides. Therefore, if you have timed your presentation so that it ends exactly on time, the slightest deviation from your carefully rehearsed path will jeopardise your chances of finishing on time. The well rehearsed, more relaxed, more confident presenter often loses time control, becomes verbose, or launches into an unrehearsed digression with the best of intentions. Sometimes, the presenter, under the spell of an attentive audience, basks and lingers in the pleasure procured by his or her oratorical talent. Other times, the presenter, eager to be understood, over-explains or lectures. It is easy to say, *it will never happen to me*. It happens, and by the time you realise it, it is too late. You will have to adjust in the worst possible way: by rushing through some slides, or by skipping them in full view of the audience while uttering some lame excuse such as *this slide is not very important*, or *in the interest of time I will skip this slide*. Such unfortunate happenings are avoided if the presenter designs for flexibility.

There are four major ways to design for flexibility: 1) give yourself extra time; 2) prepare two presentations, a long one and a short one; 3) prepare escape routes; or 4) divide your content into homogeneous segments ordered according to some principle such as from general to specific, or from simple to complicated.

Give yourself extra time. You can do it the wrong way. Unrehearsed presenters are so nervous that they rush through their slides, speaking faster than needed. They unintentionally skip here and there or parts of the slide they intended to explain, and finish before their allotted time (sometimes well before). As a result, the audience understands a fraction of what would have been understood had the presenter rehearsed enough. The right way to give yourself time is to rehearse and thus reduce nervousness. Since less is more, you have tailored your presentation to be completed one or a few minutes before the allotted time. That way, you are sure to finish on time if something happens to slow you down. Because you know time is no longer a rigid tyrannical master, you are less tense.

Prepare two presentations. It may seem unnecessary and terribly costly in terms of preparation time, yet the writer does it routinely. The abstract of a paper is a short paper. Budding entrepreneurs learn that they have to be ready to explain their idea to the hurried venture capitalist in the time it takes for an elevator to go up to the top floor of a high-rise. There is always a way (as long as there is a will) to present your scientific contribution in less time. Are you ready to say *yes* to the session chair who asks you to cut down your presentation time for reasons beyond his or her control? Will you get your plane when you are the last presenter on the last day and you have an evening flight to catch at an airport that is 20 miles outside of town on a Friday evening at peak traffic time? Are you willing to trade formal presentation time against question and answer time? Are you willing to take the time to prepare two presentations, one long and one short?

Actually, the preparation time for two presentations may only be slightly longer than for one. The shorter presentation may have most of its slides in common with the longer one. What changes is the level of detail you cover, like the abstract of a paper vs. the whole paper. Naturally, as in an abstract, the key result (or contribution) is given with the level of precision expected by the audience. Making it shorter does not mean making it more general. You still have to connect your contents with the title and fulfil the expectations of the audience (an audience expects more from a long talk than from a short one).

Prepare escape routes. Escape routes are short cuts that are invisible to the audience. They allow you to skip a slide without the audience's knowledge. As far as people know, you have been down Main Street from beginning to end. An escape route is only one click or a few key punches away. But before discovering how to take a short cut, you must carefully look at your slides. Some slides cannot be skipped: the title, the hook, and the conclusion slides for example. Any slide mentioned in the conclusion is also immovable because a scientific audience dislikes being asked to accept at face value conclusions that have not been backed up with evidence. All other slides, the "skip-me" slides, can be skipped if you are out of time.

To identify the slides you could skip, go into the slide sorter view (PowerPoint) or the light table view (Keynote). The default value for the miniature slides is usually too small (*66%* in the slide sorter for PowerPoint, and *small* for Keynote). You want it at 100%. Therefore, select all the slides (rubber band or select all). PowerPoint puts the command under the *View* menu. Select *view/zoom...* and then click on *100%*. Keynote does not place that command under a menu. Instead, hold the menu at the bottom left of the window, and select *large* (see Fig. 6.29).

You now see all your slides, and it is easier to plan your escape route. Identify the slides you could skip when pressed for time. An escape route starts on the slide immediately preceding the skip-me slide(s), and ends on the next permanent slide.

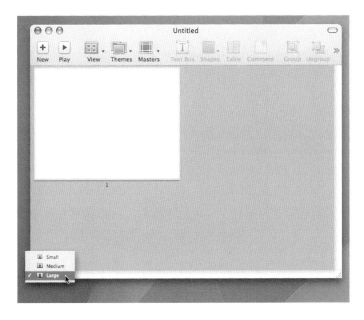

Fig. 6.29. Keynote [View/light table] slide size setting dialog box.

How does one create an escape route? There are two ways: the keypunch and the button. The keypunch consists in typing a sequence of keys on your keyboard to move transparently to the next slide in your talk. For that, you need to know the number of the next permanent slide from where you are. Naturally, next to you on the lectern is a print-out of your miniature slides with their numbers. Discretely look at it, identify the number of the next permanent slide, type that number on the keyboard, and as you press return or enter (do not forget to do that or nothing happens), the software takes you to the next permanent slide. If you are away from the computer, you need to return to its keyboard. The other way of creating a route is with an invisible button hyperlinked to the next permanent slide in your presentation. Again, you need to be next to your mouse to click on that button (unless your presentation remote includes a mouse).

How do I create a transparent button, you may ask. In PowerPoint, follow the menu sequence *slide show/action buttons/custom*. Your cursor changes into a cross. Drag a rectangle over a blank part of the screen (chose the same location for all escape route buttons to make your life easier). Depending on your default values, it is most probable that the shape will look like that of Fig. 6.30.

As soon as the button is drawn, a menu opens (at least on the Mac it does), and you are asked what you wish to hyperlink this button to. Follow the sequence *Hyperlink to:/slide...* click on the permanent slide you wish to escape to, and click *OK* to exit the dialog box.

You want the button to be transparent; therefore, the colour and the borders have to go. For that, double click on the button, or control click on it if you have a Mac (or right mouse click if you have a PC) and then select *format autoshape*.

Select *Colour and lines/Fill/colour/no fill* and *Line/Colour/no line*. Your button turns invisible. But fear not because while running your slide show,

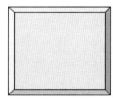

Fig. 6.30. PowerPoint typical custom action button.

Design for Presenter Flexibility 141

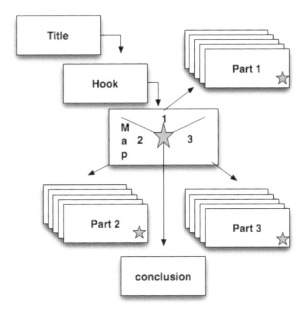

Fig. 6.31. Scheme for navigating through a large number of slides transparently. The small stars represent buttons hyperlinked to the map slide. The large star on the map slide is hyperlinked to the conclusion slide.

your cursor changes from an arrow to an index finger as soon as the mouse goes over the transparent button. A simple click and the software takes you to the next permanent slide.

The last way of creating escape routes is through dividing your content into homogeneous ordered segments and use a slide mentioned in an earlier chapter, the Map slide. Schematically, the principle works as per the following diagram (Fig. 6.31).

 Watch the "Map slide" video on the DVD

This presentation is actually the sum of three mini presentations, each with its introduction (hook) and even summary. After the title and the main hook, comes the map slide. In this example, the map is divided into three segments numbered one to three. Each segment is hyperlinked to the first slide of the mini presentation. At the end of each mini presentation, the last slide is hyperlinked to the map slide. The centre of the map slide, here characterised by a grey star, is hyperlinked to the conclusion slide.

Each slide in the mini presentation has an invisible button (here symbolised by the grey star) located at the same place on each slide for consistency and ease of retrieval (the button can be visible: the company logo that is on all your slides could be used as a button). These buttons are hyperlinked to the map slide. This way, the presenter can at any moment click on these escape route buttons and move back to the map slide without the audience noticing slides are skipped. It is advisable to order the slides in the mini presentations, for example, from the more introductory to the more specific and technical. The presenter can then easily tailor a presentation to the knowledge level of the audience and skip the slides that are too technical.

It may happen that you are given more time, not less. Are you prepared for such an occasion? Add slides (usually technical but not necessarily so) after your conclusion slide to answer more questions during a longer Q&A session. You cannot access these extra slides transparently from your main presentation unless you built links to them; therefore, link your backup slides to your main presentation, and if time allows, you can decide to present them too. Just make sure they are still part of your story.

If you jump to a hyperlinked slide that is outside your main presentation, you need to return to the main presentation afterwards. There are two ways of returning to the right place. The first way is interesting but fraught with problems (Vladimir experienced some of these). A better way is to keep your extra slides inside your PowerPoint file and arrange to return from the hyperlinked slide to the rest of your story by placing an action button on the hyperlinked slide itself. The action button dialog (Fig. 6.32) gives you the possibility to return to the *last slide viewed* (do not choose *previous slide* because that is the slide located before the one displayed).

In some cases, however, you may want to return to the slide right after the last slide viewed. In this case, you need to change the settings in the dialog box that opens when you draw the button shape from *last slide viewed* to *slide…* and then choose your slide (Fig. 6.33).

In Keynote, you will find everything you need under the Inspector Hyperlink tab (the white return arrow in a blue circle, close to the QuickTime logo). Creating transparent buttons in Keynote follows a similar process: create a shape, then in the Inspector window, click on the graphic tab and edit the shape (*fill – none; line – none*). In the process, you have to admire the talents of the human interface team at Apple. In Keynote, the totally

Design for Presenter Flexibility   143

Fig. 6.32. PowerPoint [Slide Show/Action buttons] setting dialog box.

Fig. 6.33. PowerPoint [Slide Show/Action settings/Slide...] setting dialog box.

Fig. 6.34. Keynote's way of indicating the presence of a hyperlink on a slide.

transparent button keeps a faint grey outline, and as soon as you hyperlink it to a slide, a little Hyperlink Logo (Fig. 6.34) shows inside the grey outline (but on the projected slide, it is totally invisible of course).

In PowerPoint, when the button becomes transparent, you no longer see it and unless you know where it is, you waste time finding it.

Design for Persuasion

What is the audience to be persuaded of? The audience wants to be persuaded that you are an expert in your field so that they can learn from you and trust your results. But let's consider your wishes also. What do you, the author, want to persuade the audience of? You want to persuade the audience that you are an expert whose contribution to science is relevant to people's needs. If nobody uses your work or at least evaluates it, you have accomplished nothing. Naturally, much persuasion will come from you, and in particular your confidence and the ease with which you present your (well-rehearsed) material. Indeed, the synergy between your slides and your delivery is great. In this section, although our purpose is to consider how your slides strengthen or reduce people's trust in you, it will not be possible to dissociate your slides from the way you deliver their contents to the audience since the two are intimately intertwined.

How do you persuade the audience you are an expert? It is not possible to provide enough data on your slides to satisfy the experts in the audience. Complex formulas, large tables, detailed diagrams take much time to explain and time is precisely what you are short of; therefore, your evidence can only be partial. However, people can still be persuaded based on partial evidence if they trust you are an expert. But if they do not know you, if you are not yet a highly published researcher in your field, the only thing the audience can rely on to judge your expertise is your presentation. Through you and your slides, the audience will get subtle clues that consolidate their opinion of you as an expert: your intellectual honesty, your clarity of expression, a non-hesitant speech, the clarity of your slides, and finally, your calm, accurate, precise answers.

How do you persuade the audience that they should delve deeper into your contribution after your talk? If you need to persuade scientists that your work is useful to them, you need to persuade yourself first. Do you have that level of confidence? Could you identify how your work would be useful? Are you ready to mention how in your introductory or concluding statements?

Demonstrate Intellectual Honesty

Persuading a scientist is persuading someone whose certainties are constantly assailed by doubts. Were the data measured accurately? Were the samples representative of the population studied, without bias? Were the modelling assumptions justified? What is the impact of the complexity-reduction choices on the applicability of the findings to real world problems? This questioning about limitations is healthy and symbolic of intellectual honesty. A presenter perceived as intellectually honest is by definition more believable and convincing than a presenter who is perceived as having results too perfect to be honest, claims too broad to be true, or contributions too extensive to be his or hers alone.

Mention your sources. It often happens that someone else's visual helps you strengthen or visualise a point. When you do use such a figure, write down the name of the source next to the visual. The author or someone who knows the author whose work you display may be in the room. Not writing down the source is the same as claiming the visual is yours and this is a breach in intellectual honesty (besides being a breach in copyright law). It is appropriate to repeat the name of the source in your speech in two cases:

1) When, for lack of space on your slide, the name of the source is written in small letters unreadable from the back of the room.

2) When the name of the source contributes to making your findings even more authoritative. To be authoritative, however, you do need to trace the visual back to its original source. Many visuals available on the web are copied from other websites.

Mention your hypotheses, your assumptions, and the reasons for making them. Your assumption is justified because it reduces the computational load or because it makes the problem more tractable (for example, by reducing its dimensions). Show that it is a reasonable assumption. For example, indicate that the factor ignored is negligible, or that the experimental results are close to the simulated results. In other words, defend your choices during your talk, not during the Q&A when being on the defensive forces you to be negative about your work and say, *we did not* instead of *we did*. Hypotheses and your rationale for them can form a part of the story of your contribution.

Acknowledge other people's work. Unless your contribution is unique and does not depend on anyone else's work (a rare case), it is always appropriate to mention those on whose shoulders you climbed to look at the world and make your hypotheses. Often, presenters avoid mentioning other people's work when comparing. Such comparisons are limited to a few factors where their work appears to do better than the rest. If that is your case, justify why the factors used for the comparison are important. As is often the case, because space on the slide is restricted (for legibility's sake), comparisons are made using adjectives or plain numbers. And since each adjective is an undemonstrated claim, it is always questionable. If the authors of these methods are attending your talk, chances are they will challenge you to justify yourself during the Q&A. Therefore, be prepared to justify these adjectives or numbers with solid evidence on more detailed slides following your conclusion slide. If you do not have the details, do not compare. Do not expose yourself to ridicule during the Q&A session, thus losing your authority in front of the audience.

Demonstrate Confident Expertise without Crowding the Slides

Use specific words instead of general words. Detecting a non-expert is trivial for an expert. Non-experts cannot be precise, they use many words and adjectives instead of informative visuals, and they keep to general statements. Non-experts are rarely clear when they explain complex diagrams. On the other hand, experts are tempted to show their expertise by crowding each slide with many detailed points, or by re-using complex graphs and tables from their paper without simplification. Judging from the reaction of audiences subjected to such presentations, it may work. People tend to attribute their inability to understand a presentation to their own lack of expertise, therefore viewing the presenter as an expert… albeit one unable to express ideas clearly. Since you are reading this book, I presume you do not wish to be that kind of "expert". Instead of drowning the audience under a deluge of arguments, major and minor (as if the accumulation of facts convinces more than their relative worth), use the most salient argument to convince. This often entails simplifying a graph or extracting the critical steps from a complex series of steps.

Display confidence during the talk and especially during the Q&A. The Q&A session can rescue a presenter whose formal presentation was poor. This happens when the presenter displays a deep knowledge in both understanding the questions and responding to them accurately and with great calm and confidence. Anxiety during Q&A is often seen as an attribute of non-experts. Experts are more at ease with one-on-one questions within their field of expertise.

Describe Your Data Clearly in an Appropriate Graph

Numbers have a greater convincing power; that is why they are used in visuals to help substantiate a text claim. However, numbers have many sources: some are raw data straight out of data collection, some have been processed to remove obvious errors in data entry, some come from experiments, some are generated by the equation of a model, some are averages over many observations, some have been normalised, some are rounded, some are discrete, some are continuous… the list is long. Many times, the audience is unaware of the source and nature of the data. The audience may even imagine data come from experiments when in reality they come from simulations. The audience sees a curve and imagines continuity when in fact the data points are discrete. To avoid such misunderstandings, it is essential to make sure the choice of curve matches the type of data, and to describe how data was obtained when presenting a graphic or a visual. It takes a little more time, but it avoids troublesome questions during the Q&A.

State Your Point Clearly in a One- to Two-Line Heading

Clarity of claim is necessary for persuasion. You cannot be persuaded by what is hard to understand or what is expressed confusedly. To make sure the audience gets your story points, write each one as a full sentence in its own dedicated space at the head of each slide. Underneath the heading, which in essence represents your claim, provide visual evidence (at most two visuals — for comparative use for example, but preferably one visual only). That format is simple, and extremely effective. It prevents you from trying to make many points on one slide, and it enhances clarity by privileging the role of visuals.

When you create a slide, do not only work on its contents, decide also on the order in which you will reveal or explain these contents. The font size, the colours, the arrows, the layers, what is on top, at the bottom, in the centre, on the left side, on the right side of the slide ... Prepare the contents so that the eyes scan them in a movement that you have choreographed. Your words simply accompany the visual path. You should know what people look at first when first viewing your slide. If you do not know and cannot imagine it given the slide layout, bring people in front of your slides prior to the presentation and ask them in what order they looked at the elements on your slide. After doing this a few times, you will not need to guess, you will know. A well-designed slide guides all members of an audience along the same visual path.

Part IV

The Presenter

Executive Summary

So far, the presenter has only been involved in content selection and slide creation. Now comes the time to face the public and "deliver" the slide contents on a white screen, using a panoply of tools: Tools that guide the attention of the audience, such as laser pointers or voice clues; tools that extend the presenter's reach, such as microphones and presentation remotes; tools that bring knowledge to life, such as PowerPoint or Keynote. Through the miracle of communication, words, voice and gestures, knowledge becomes food for thoughts.

The presenter is the host, and the scientists are the guests. Good hosts are prepared. They are attentive to the needs of their guests. They are in plain sight, seen by all, and accessible. They are hospitable. They speak intelligibly with confidence and authority. They also speak with interest to encourage the audience to evaluate and appreciate their scientific contribution.

Good hosts consider each guest as worthy of their care and attention. They know how to gracefully handle difficult guests during the Q&A session, the ramblers, the hijackers, the scientists whose corrupted English is defying comprehension, and even the guests eager to disagree. Prior to the presentation, good hosts prepare answers to potential questions. They know how to listen to, rephrase and answer each question, be it non-sensical, difficult, controversial, or hostile. They mastered the techniques to directly bring to the screen a visual for fast answer support.

7

THE MASTER OF TOOLS

Screen, Pointers, Mikes, and Lectern

The Projection Screen and Lectern

Where do you stand in relation to your screen? The logical answer is to stand behind the lectern: it is where the computer and microphone are located, and it is where the organisers of the meeting expect you to be. That is why the majority of people stand there like amputated Greek statues whose only body part you see is the bust.

Observe Fig. 7.1. Should you stand at location 1, 2, 3, or 4? Does it matter? What are the pros and cons for each location?

Location 2, behind the lectern, is indeed the expected location. For the faint-hearted, the lectern is a bastion behind which to find protection against scientists who have mastered the art of war, a wooden fortress against which to lean and cling to while facing the assailants. The lectern harbours a powerful microphone, but it is a lectern mike. This means that each time you turn toward the screen and continue speaking, the audience feels like reaching out for a remote to increase the volume ... but of course, there is no remote. The screen of the computer is conveniently placed on the lectern table to enable you to keep an eye on the audience and the other eye on your computer display. But space on the lectern table is limited; its narrow flat surface is more like a precarious ledge on which too many things stand: your notes, possibly a glass of water, the computer screen, the mouse, and possibly a pen and a clock. Standing behind a lectern does not do justice to the nice clothes you are wearing. You might as well be wearing shorts and sandals, because

Fig. 7.1. Possible presenter locations on the podium.

the audience will only see the upper part of your body, and if you happen to be short, the microphone may even hide part of your face. Tall people should not rejoice at this time. Lectern mikes always seem too short for them. They force them to bend forward while pulling the neck back, a posture that impacts both breathing and voice clarity and is not sustainable over a long period. Therefore, I give this location a score of 5 out of 10.

Location 3, centre stage, looks good (on paper). You are facing the audience and are at the centre of attention. Of course, you may have to turn your head left and right to make sure the whole audience benefits from your attention, but that position gives you the ability to move to the left and to the right freely. It also requires that you hold a wireless microphone or wear a clip-on wireless Lavaliere microphone; both give you more flexibility than the fixed lectern mike, or a corded mike. It would be the perfect position and score 10 out of 10 if the screen was high above you, and you never needed to point to anything on the screen. Alas, often the screen is low and the projector beam traverses the centre of the podium and drapes you with the contents of the slide. Therefore, if you want to stand there to address the audience personally without slide support, you have to temporarily blank the screen behind you. This is possible while being in slide show mode. Simply press the "B" key on your keyboard for PowerPoint (or Keynote) to temporarily blank the screen. If you have a presentation remote, press the button with a black square. It acts as a "B" key. Furthermore, if you really need to use your laser pointer, you have to move to the side of the screen, otherwise you would turn your back on the audience. A better option would be to have all the pointing done within PowerPoint or Keynote (arrows, callout boxes). This location is only recommended when the screen is blank or high up above you. It deserves a score of 8 or 0 depending on the situation.

Location 1, stage right, as well as location 4 are on opposite sides of the podium while the screen is in the centre. Both locations enable the whole body to be seen, which is a better way to relate to the audience. Both locations require a mobile microphone. Both locations avoid the projector beam. Which position is better? Location 1 is closer to the computer. If you do not wish to use a remote to move to the next slide and prefer to tap the spacebar or click the mouse, position 1 lets you do that. However, the lectern and the sides of the podium limit your range of movement. The presentation pundits claim that the presenter should always stand to the left of the screen (from the audience's viewpoint), in location 1. They say that people read from left to right and therefore, when they finish reading they return to their base position, the left of the screen (stage right from the perspective of the presenter facing the audience). They also say that this location allows you to continue facing the audience even when you turn sideways to point to something on the screen. Alas, location 1 is only good for left-handed presenters who handle the laser pointer in their left hand. Otherwise, if you attempt to use the pointer with your right hand, you have to awkwardly bring your right arm across your chest and twist your neck and turn sideways to take better aim. This position scores 8 out of 10 for presenters not using a presentation remote.

Location 4, stage left, does not limit the range of movement as much as location 1. You can point to the screen with a laser pointer using your right hand. The disadvantage is that you need a presentation remote or you need someone familiar with your presentation to change the slides for you. Moreover, you would have to cross the stage to access the computer mouse unless you use a presentation remote device that also features a mouse button such as the Keyspan presentation remote. Crossing the stage means that you have to blank the screen prior to crossing. Doing this once or twice is fine. And doing it well requires a little bit of choreography: from position 4, you would blank the screen, move centre stage, engage the audience for 20 seconds, move to the computer, launch an animation, stay in position 1 for a while, then blank the screen again and move to position 4 prior to starting a next segment of your talk. The pundits do not like this position because it goes against reading direction. But their opinion is only valid when there is much text on the slides (there should not be much). Resting your eyes on the presenter standing on the audience's right after reading a line is just as easy as doing a mechanical carriage return with the eyes. This position also deserves an 8 out of 10 because it enables you to move and be seen, and to focus audience attention on you.

Now that you know the strengths and limitations of the various locations, work around these limitations to score a 10. It is possible but it requires choreography and many rehearsals. Watch America's ex-Vice President Al Gore or Apple's CEO Steve Jobs during their talks. Their movements are carefully planned. Never miss an opportunity to escape from the magnetic attraction of the lectern. Your unconventional departure from the norm conveys the message that you intend to face people and personally communicate to them.

Laser Pointers and Presentation Remotes

 Watch the "Presentation remotes" DVD Movie to discover the pros and cons of four popular models

Presentation remotes free you from having to be close to the keyboard or mouse. They enable you to move outside the immediate perimeter of the lectern and computer. However, their main drawback is that they occupy one hand, usually the right hand, the one used for expressive gestures. Your gestures become unnatural. Furthermore, you only have one hand left (the left hand) to hold other objects during the presentation: your glasses, a glass of water, a microphone, a marker pen to write on a flip chart, a microphone, or a laser pointer. Juggling with one object using two hands is easy; juggling with many objects makes life needlessly complex.

Presentation remotes are all different in terms of functionality and ergonomics although they all have common features such as the ones shown in Fig. 7.2. They all have pluses and minuses. None are perfect. Choose the one that seems best for you and your budget. I do recommend, however, that if presentations are taking a significant amount of your time in your line of work, you purchase your own presentation remote and carry it with you to use in your presentations. You will be more comfortable with it and make

Fig. 7.2. Face of Kensington presentation remote. Top button is for laser. Grey square is for blanking screen. Left arrow is for returning to previous slide. Right arrow is for going to next slide. Face is slightly depressed allowing thumb to find resting position at the centre.

fewer mistakes, but since their user interface is not perfect, don't expect to be perfect either.

Presentation remotes are wireless. Their radio signal enables you to stand ten metres away from the computer (or more). They do not need to be pointed toward the computer or toward the screen to work. Pointing to the screen to move to the next slide is a common mistake presenters make.

The right-arrow key (next-slide) is going to be the key the most used on your remote. It is the equivalent of a mouse click, or pressing the return key, the right arrow key or the spacebar on your keyboard. Most remotes place that key on the right of the left-arrow key, as it should be. This key should be pressed briefly. On some remotes, when you keep the key pressed for a few seconds, it acts as a fast forward function. You certainly do not want that in a normal presentation, so press that arrow briefly only, as if you clicked on your mouse button.

The left-arrow key (previous slide) should never be pressed. It is there only for people who make a mistake and press the next slide key instead of the adjacent laser beam button. A presentation is a one-way walk through the slides. If during your talk you need to return to a previously shown slide, it is because your slides do not stand alone. Each slide should only depend on its contents to be understood.

The laser beam button is usually in between the left and right arrow keys (a user interface blunder). Using a laser beam as a pointer has two advantages and countless drawbacks. On a dark screen, the red or green razor sharp dot really catches the eye. The bright dot does shepherd the eyes of the audience to a rallying place better than a sheepdog, and here ends the short list of advantages. Now on to the long list of problems:

1) It eats up battery power. Of course, you only find out on the day of your presentation that you need new rare-to-find-button batteries to get a reasonably bright spot that people can see without plunging the room in total darkness.

2) Stand-alone laser pointers with brighter and adjustable red dots can also be found. They are independent of the presentation remote. But if you also use a remote, your two hands are both holding something, making it impossible to have natural looking gestures.

3) It does not work well on a white background, such as the background recommended in this book.

4) It betrays your fear and lack of confidence as efficiently as a quivering voice by amplifying your trembling hand movements.

5) Since most presentation remotes place the laser beam button next to the right arrow key, you routinely confuse one with the other, thus creating havoc during the talk.

6) Laser beams are dangerous when pointed to a person's face. In some countries, using a pointer with a powerful green laser beam requires an authorisation because of its danger. It could permanently damage someone's

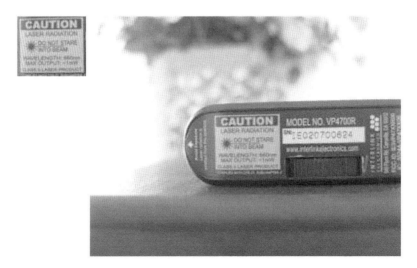

Fig. 7.3. Notice found underneath the Interlink Electronics presentation remote. *Caution. Laser radiation. Do not stare into beam.*

eyesight. Red laser beams are also dangerous but of much lower power (Fig. 7.3).

7) Since most presenters do not know how to handle a laser pointer, they keep circling a place of interest while talking. Mesmerised by the bright dot running in circles, the audience gets caught in the hypnotic "follow-the-dot" mode and becomes listening impaired.

8) Some presenters cannot aim properly (lack of gameboy practice) and they distract the audience by taking some time to hit their target, or by holding the laser pointer with both hands as if it was a loaded gun about to recoil.

Therefore, if you have a presentation remote and feel like using its laser pointer, look first at the remote to identify the laser button so as not to inadvertently click the adjacent right or left arrow buttons. Then, steady your hand as you aim. Finally, move the dot quickly to the area you want to audience to look. As soon as you get there, keep the beam on for one second and then turn it off. Do not circle, underline, shake, or use any other distracting movement unless you have a rock steady hold and great aim. Even then, only do this once (not repeatedly). Do not talk as you point the beam to the screen: let the audience focus on your beam. Then turn off the beam and talk while facing the audience.

The black square/rectangle button corresponds to the "B" key on your keyboard. It blanks the screen. Use the black square when you have to move in front of the screen to go to the other side of the podium. Use it also when you want the audience to focus on you and not on the screen so that people listen attentively to what you have to say.

The Function Key starts the slide show, but this function is often available for the PC platform only and not the Mac platform (Logitech presentation remote Fig. 7.4).

Fig. 7.4. Face of Logitech presentation remote. Centre button is for laser. Grey rectangle is for blanking screen. Left arrow is for returning to previous slide. Right arrow is for going to next slide. Function key has two functions (PC only). While in slide show mode, it functions as an escape key and terminates the slide show. While in normal view, it starts the slide show.

Fig. 7.5. Right: Logitech presentation remote. Centre: Timer button on side of remote sets time in five-minute increments. Left: close up of face revealing the timer. Black bar under time display is proportional to overall time remaining. The remote vibrates two minutes before the end of the set time reminding you that you should be on your conclusion slide. It vibrates one more time briefly at the end of the set time.

The timer/clock. Although having a timer or a clock on the remote looks like an extremely good idea, in reality, the presenter rarely looks at it. Having it function as an alert is useful. The Logitech model (Fig. 7.5) vibrates to alert you when it is time to conclude. You could also set up the alarm mode on your mobile phone and get it to vibrate at the appropriate time while putting your calls in silent video alert mode.

Microphones

 Watch the "Microphones" video on the DVD

Microphone pick-up patterns

Unless small, meeting rooms have a microphone. These mikes are usually omnidirectional microphones. They pick up the sound all around them (that includes room and audience noise); if you move to the right or the left of the mike, you will still be heard. This is a key advantage of omnidirectional mikes. These mikes are also less sensitive to the bursts of air created by your "B"s and your "P"s. When you move closer to an omnidirectional mike, apart from becoming louder, your voice colour does not change. Not all lectern mikes are omnidirectional, however.

Some microphones are cardioid mikes. They are so called because they capture sound in front of them in a lobed heart-shaped pattern. Very little sound gets in through the back of the microphone (that is just fine because that is where the audience is). The cardioid mike is more likely to create popping sounds as your lips release the "B" and "P" consonants with a gust of air that hits the surface of the mike. This is why some microphones have foam bonnets to dampen airbursts and reduce popping. When you get closer to a cardioid mike, however, your voice gains a warmer, deeper sound because the mike naturally enhances low frequencies. It is called the proximity effect.

You have just been given two ways of identifying the type of microphone you are likely to face on a lectern: 1) if you talk around the microphone (or rotate the mike as you talk into it) and the loudness of the sound remains the same, it is an omnidirectional microphone; and 2) if it pops with the plosive consonant "B" and "P" or if your voice tone changes quality as you move closer to the mike, it is a cardioid microphone.

Microphone types

The lectern microphone

Lectern mikes are usually omnidirectional microphones because they still pick up sounds when the presenter moves (slightly) to the right or to the left of the mike (Fig. 7.6). Cardioid microphones are not so forgiving. They require that you remain within their heart-shaped pick-up pattern.

Fig. 7.6. Typical lectern microphone.

Fig. 7.7. Lapel microphone with its battery pack.

The lapel microphone

Lapel mikes (also called lavaliere mikes — Fig. 7.7) are clipped on the lapel of your shirt, polo shirt, or front-buttoned blouse (or to a hand-gathered/rolled bunch of garment for all other situations). Be dressed for the occasion. When you present, wear clothes with a rigid lapel. Avoid T-shirts, sweaters, and for ladies, avoid scoop neck blouses and dresses without pockets. Avoid wearing a windbreaker jacket during your presentation, even in a cold room. There is not much wind there. A presenter needs to be "presentable". And do not clip your lapel mike to anything that can move while you speak, such as the cord holding your conference badge or to the side of an open front-button sweater, because as you move, these move also and the tip of your mike rubs noisily against your clothes. The clip-on mike should be placed one stretched hand away down from your chin, as you are looking straight ahead. Do not adjust the height of the mike higher, even if your voice is soft. Let the audio technician adjust the volume on the meeting room mixer/amplifier.

A lapel-mike does not pop simply because your mouth is above it. Even though some lapel-mikes are wired directly to an audio plug in the wall (or the mixer), most are wireless. This means they are attached to a transmitter pack. Sometimes the pack is part of the mike, sometimes it is detached. On the back of the detached pack, the manufacturer often provides a clip to hang the pack to your belt or on the rim of your trouser/skirt (out of sight behind your back). Alternatively, you could pocket the pack (as long as you have a

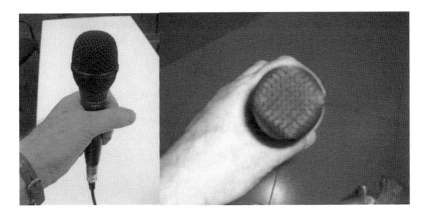

Fig. 7.8. Two hand-held cardioid microphones. Vertical (left) and horizontal (right) hold.

pocket). Therefore, dress appropriately: wear a trouser with a belt or pocket, or a skirt to which the pack is easily clipped or pocketed. Do not find yourself having to use one hand to hold the power pack and the other hand to hold the presentation remote.

The hand-held wired or wireless microphone

Be aware that there are two kinds of mikes: those you hold vertically with the microphone side carrying the company logo facing your mouth, and those you hold horizontally (like singers) with the tip of the mike pointing toward your mouth (Fig. 7.8). If you do not know which kind of mike you have, try both positions and listen to the difference in volume. The volume of a singer's mike held vertically is much lower than if held horizontally.

Wireless hand-held mikes usually have an on/off switch and require a power/transmitting pack which is at the bottom of the mike, inside the body (Fig. 7.9). The quality of the sound of wireless mikes is a function of the battery charge. It is best to make sure your battery is new before you start. The technician always has a replacement battery available. When you hear crackling sounds, it is time to change the battery. And when you hear no sound, it is time to check if there is a battery, if the battery is dead, or if your microphone is turned on!

Microphone use

Sound intensity (measured in decibels) drops by 6 decibels (half the perceived loudness) every time the distance from your mouth to the mike doubles. Is

Fig. 7.9. Wireless hand-held microphones with on/off switch.

this important? Very much so, and here is why:

1) If you stand far away from (or very close to) your mike in the first minute of your talk, the technician will compensate by raising or lowering the volume on the sound mixer, hoping you will continue to speak at that distance. If you then move closer to or farther from the mike during your talk, the loudness of your voice will change and, as far as the audience is concerned, your voice will fade in and out, an unpleasant ear-straining experience. Do not rely on the technician to move the mixer levers up and down to keep your voice loudness constant. He may no longer be in your room, or he may be dozing off waiting for the end of your talk to come back to life. The junior presenter (or the tall presenter) tries to adjust the loudness at the beginning of a talk by bending the body forward toward the sensitive lectern microphone instead of letting the technician adjust the volume on the mixing console. Two things then happen: the voice loses its strength because the air passage is bent "out of shape", and the body starts cramping after a while. To avoid this, adjust the position of the gooseneck of the lectern mike so that you can stand normally. If you are tall, keep it straight up pointing toward your mouth and if that does not work, ask for another mike. Do not try to hold the lectern mike while you talk.

2) When you turn toward the screen away from the mike, the sound that reaches the microphone is the sound that bounced off the surface of the screen. If you are wondering why the audience can't hear you anymore,

simply consider how far the screen is, multiply that distance by two (the sound that bounces off the screen needs to come back to your mike), and then compare that distance with the normal speaking distance between your mouth and your mike. You will discover that the audio volume has decreased by 4 to 5 times in less than a second. As you turn back toward your mike, the sound suddenly increases — a most unpleasant experience for the audience. Therefore, speak toward the mike, keeping mouth-to-mike distance as constant as possible. This is true for all mikes. If you want to know more, find out about the "Inverse Square Law". It states that direct sound levels increase or decrease by an amount directly proportional to the square of the change in the distance between your mouth and your mike.

Testing whether the mike is on, is NOT done by tapping on the mike. Snap your fingers close to the mike and listen. Whether the mike is "live" or not is immediately clear. There is no need to ask the audience whether your mike is on or not. Wireless mikes have an on-off switch. Make sure you know where it is. The previous speaker may have turned the microphone off. If you hear no sound even after turning on the switch, look for a spare battery, or in the direction of the audio technician and gesture to him. DO NOT start your talk without audio, wait until the problem is resolved and help resolving it. DO NOT think for one moment that, because your voice is loud, you do not need a mike. The people at the back of the conference room will be sure to let you know!

Turn off your mike before handing it over to the next speaker. If you are given one for your presentation, keep it turned off while you are clipping it to your clothes, and only turn it on when ready to use it. As you reduce the distance between a hand-held cardioid mike and your mouth, you will notice a significant change in the quality of your voice, and hear popping with "Ts" and "Ps". Some cardioid mikes pop easily, others don't. If you hear popping, speak above your mike so that the gusts of air from your breath pass over the mike, not right into it. If you hear ear-shattering screeching audio feedback, move away from where you are to where you were. It means that you are standing too close to a loudspeaker or in front of it (you should at least be 10 feet away). If you hear some feedback wherever you are on the podium, it means that the fader that controls the volume of your mike on the sound mixer is set too high. It may be that the previous speaker had a soft voice and the technician did not adjust the volume after his talk. Ask the audio technician to reduce your mike volume.

Audio and Light Control

Are you surprised to see such a large heading over such a short paragraph? It is your duty as a presenter to proactively seek to know how light and audio are controlled in your meeting room. Indeed, it is your duty to discover the room prior to the start of your presentation. You will probably go there to meet the chairperson of your session and install your presentation on the lectern computer. You are in charge, even though there may be a technician in the room. You are the presenter. You control everything that may affect the quality of your presentation, and that certainly includes the control of light and sound. If some of your slides, for readability's sake, require the lights to be brought down significantly during your talk, arrange this prior to the start of your presentation. Talk to the technician, or arrange for someone to stand by a light switch. If you have a video segment containing audio on a slide, your presentation computer needs to be connected to a line input. Do not wait until during the presentation to discover that you have no audio out on the computer. The technician would have been more than willing to provide audio out. Even if the computer audio out is connected to the sound mixer (Fig. 7.10), you may have failed to inform the technician that you had audio needs; as a result, the fader on his sound mixer remained down.

Indicating your audio needs prior to the presentation is critical, and finding a way to resolve the difficulties is your responsibility. For example, if there is no provision for audio out, you may want to place your microphone close to the computer speakers. YOU ARE THE ONE IN CHARGE, no one else.

Presentation Software (Keynote and PowerPoint)

You have probably seen presenters stuck on the first slide of their presentation because they do not know how to start a slide show. Other people prepared their presentation. How did you feel when that happened? Irritated? Did the presenter lose your respect from that moment on? Unless one moved from Windows XP to Vista, or from one version of the presentation software to

Fig. 7.10. Small sound mixer with white faders controlling the volume of six sound inputs.

another with a vastly different user interface, this situation is hard to justify. Furthermore, it demonstrates that the presenter has not rehearsed (never a good sign). A presenter should be familiar with one or several presentation programs, including some image manipulation packages such as Photoshop to enhance the readability of some visuals or even text. At the very least, a presenter should know how to do the following tasks:

1) **Hyperlink an image to a slide, and remove a hyperlink.**

(See **PowerPoint** help: *Remove a hyperlink* and *Create a hyperlink*; Hint: right click on graphic or image. For the Mac users, on a one-button mouse, the equivalent of a right click is control-click.)
(See **Keynote** help: *Using hyperlink to* …. The help mentions how to stop a slideshow, but you can do many other things as well. To remove a hyperlink, just select the object to which a blue circle and white arrow is attached. Click inspector in the toolbar, click the hyperlink inspector button, and then deselect *Enable as a hyperlink*.)

2) **Make a transparent button and hyperlink the button to another slide or to another presentation.**

(See **PowerPoint** help menu: *About hyperlinks and action buttons*, and *Insert an action button*. Hint: to make the button transparent, double-click on the button and choose *no fill* from the fill colour pop-up menu and *no line* for the line colour pop-up menu. It is better to hyperlink the button before making it invisible, because it really becomes invisible and unless you know where it is, you may find it difficult to find in a slide with many objects. If you have linked an object to another PowerPoint presentation, when that presentation ends, you transparently return to the slide in your main presentation from which you linked out. This is great!)

(See **Keynote** help menu: *Linking to a slide*, *Linking to a keynote file*, *Using shapes*, *Change an object's fill colour*. Hint: to make the shape transparent, select it, click inspector in the toolbar, then click the graphic inspector (circle over a square). Choose *none* from the *Fill* pop-up menu, and *none* from the *stroke* pop-up menu. Keynote kindly provides a grey outline around the object shape to locate it when it becomes invisible (slide show mode). When you hyperlink the shape to a slide, contrary to PowerPoint, Keynote also reminds you visually that the transparent shape is hyperlinked via a blue circle including a white arrow. Great user interface design. By default, Keynote places the shape in the middle of the screen. To directly place the shape over the text to hyperlink, press the option key as you select your shape. After selecting, drag the crosshair pointer over your text.

If you have linked an object to another external Keynote presentation, unlike PowerPoint, you DO NOT automatically return to the slide in your main presentation from which you linked out. You therefore have to create a hyperlink within that external presentation to return to the original one. Unfortunately, when you return, Keynote takes you to the first slide of your original presentation. You therefore have to remember which slide you need to go back to, and type its number. This is definitely not as transparent as PowerPoint. It is better to keep all the slides within your one Keynote presentation and to copy the slides from other presentations into it.)

3) Insert and play movie on a slide.

(See **PowerPoint** help menu: *Insert movie on a slide*. Do not forget to take the movie alongside your presentation on your USB drive if you transfer it to another computer… otherwise, you will see the poster frame of the movie and the dreaded message *The movie file "…" cannot be found. Without this file, the movie cannot play properly*.)

(See **Keynote** help menu: *Adding a movie*. If you present on somebody else's computer, it is recommended to save your movie within your presentation.

For that, you need to ensure that the *copy audio and movies into document* is selected. You find that option under the *advanced option* triangle in the dialogue that appears when you save your keynote presentation.)

4) Create an image from grouped items (image and text).

 Watch the "Scale groups of objects/images" video on the DVD

(See **PowerPoint** help: *Group objects*; Hint: Right click on grouped objects and select *save as image*; choose the png format.)

(See **Keynote** help: *Grouping and ungrouping objects*; Keynote does not enable you to easily create an image from grouped items from within the application. You can do it, though, with the help of Apple's Preview application. Simply select your group on the Keynote slide and copy it. It will go into the clipboard. Open Apple's Preview application with the menu option *New from clipboard*. Then click on the *Select* button, copy, and then paste the image back into Keynote. This is now a scalable image.)

5) Create an image from a slide.

(See **PowerPoint** help: *Save a slide as a picture*.)

(See **Keynote** help: *slides as image files*.)

6) Animate bullets one at a time.

(See **PowerPoint** help: *About animation* and *Animate text and objects*.)

(See **Keynote** help: *Moving objects on or off slides using build effects*.)

7) Remove the background (set on Master) from a normal slide.

(See **PowerPoint** help: *Hide the background graphic on a slide*.)

(See **Keynote** help: *Changing a slide's layout*. Keynote uses themes. You can create your own themes, one with your organisation's logo, and one without. Within a theme, you can have many possible layouts.)

8) Make a black slide.

(See **PowerPoint** help: *Change a slide background colour*.)

(See **Keynote** help: *Changing a slide's layout.* The easiest way to create a black slide in Keynote is by creating a new slide and applying the *black* theme to it.)

9) **Simulate the effect of the "B" key by fading to black and then fading from black, using a black slide sandwiched between two slides.**

 Watch the "B Key" video on the DVD

(See **PowerPoint** help: *Keys for running a slide show*; and *Add transitions to a slide show.*)

(See **Keynote** help: *Adding transitions between slides.*)

10) **Improve image contrast and brightness.**

(See **PowerPoint** help: *Picture formatting options*; Hint: double-click on picture or right-click and choose *format picture.*)

(See **Keynote** help: *Changing an image's brightness, contrast, and other settings.*)

11) **Use motion path or actions for emphasis or illustration.**

 Watch the "Animations" video on the DVD

(See **PowerPoint** help: *Motion path*; Hint: select motion path in custom animations.)

(See **Keynote** help: *Animating objects on slides (Action Builds).*)

8

SCIENTIST AND PERFECT HOST

Puzzling title. Aren't the organisers of the conference the hosts? Isn't the chairperson the host? The organisers are hosting the conference, and the chairperson is hosting the day sessions. But you are the one hosting your presentation. If you have hosted people, you must have experienced the tingle of excitement before their coming, the expectant glance outside the window to check for their arrival. But you would not have been filled with fear. Then why is it that people say that presenting in front of an audience is feared even more than death? Surely, this applies to irrational people who cannot control their emotions, but not to you, the logical level-headed scientist.

Somehow, you wished it were so, but fear still manages to test your confidence. This is because fear has deep roots. The first root of fear is fear itself. Fear has a way to fuel itself. It needs a starter, but once turned on, it feeds itself on your awareness of it being there. Here are six common presentation fears: 1) fear of facing an anonymous audience, 2) fear of not being able to communicate clearly, 3) fear of forgetting something important when presenting, 4) fear of being questioned, of not being questioned, and of not being able to answer a question, 5) fear of being investigated, probed, scrutinised, and possibly rebuked, and 6) fear of being rejected on the grounds of sex, race, skin colour, body height, body weight, accent, or nationality. This short list is far from complete and you could probably add one or two fears of your own. Combatting these fears is actually not as difficult as it seems. This and subsequent chapters offer ways to make these fears disappear.

Much fear disappears when you realise who you are. You are not a victim about to bleed under the razor-sharp tongues of prosecutors. You are not the accused standing without a lawyer in front of a jury determined to deliver a guilty verdict. You are not the foreigner appearing in front of an audience of linguists determined to find fault with your accent. You are not the graduate facing an unpredictable panel during your dissertation defence. You are the scientist who accepted to host scientist guests. Indeed, your name and invitation to your talk were inside the program sent to all meeting attendees.

Unless the conference has only one track attended by all, the invited guests have the choice to accept your invitation or to attend someone else's presentation. Therefore, in front of you, the audience is composed of all the people who accepted your invitation to come and benefit from your talk. They are your guests, and that makes you their host. So how does a host behave, and will people say you are a good host?

Mrs. Toldoff receives her guests

Vladimir's wife, Ruslana, had invited Vladimir's manager and a few colleagues to her home on Sunday for a blini party. The French have their crêpes, and the Russians, their blini. Of course, the Russians claim that blini are better than crêpes, and the French, don't claim anything because they have never seen a blini in their life! Blini are small thin round pancakes made with milk, egg, flour, but also yeast. The yeast makes them very light. Blini are stuffed with fishy things such as caviar, cod, smoked salmon, smoked or salted sturgeon, and with sour cream. Vladimir's blini party was popular in his neighbourhood because he made sure there was plenty of vodka to wash down the copious sour cream that coated the delicious fish. Ruslana had arranged the table where everyone would sit. Several plates, each with its towering stack of blini, were placed at regular intervals along the long table. She had dusted and cleaned the house thoroughly before the guests arrival. She was conscious of her reputation in the neighbourhood. That morning, she had even chosen the clothes Vladimir would wear. And he was too smart to argue with his wife on the day of a blini party!

The first one to arrive was Vladimir's manager. It was the first time he had come to the house and, not exactly knowing where the house was, had left home early. Ruslana heard the car door closing and was waiting at the front door, with a warm smile. At the same time as Vladimir (still in his bedroom) was shouting "Dorogaya, where is my Sunday shirt?", his manager entered and asked "Where is Vladimir?"

> "Hanging in the closet…", she shouted.
> "Really", responded the manager with a grin, "I am sorry to hear that!"

The Attentive Host

All of us remember our parents having guests. How did they host them? Chances are that prior to receiving the guests a lot of preparation took place. If it was a barbecue, someone (maybe even you) went to buy the charcoal. They bought the meat, skewers, lighting fuel, buns, beverages, ice for the drinks, paper plates, napkins, plastic cups, etc. They made sure there were enough chairs for everyone and made contingency plans in case the weather was not clement. Of course, they cleaned up the house (and themselves) and made sure the restrooms had fresh soap, towels, and toilet rolls. And at the last minute, just before the guests arrived, they checked that everything was in order and that they looked "presentable".

The first quality of a host is preparedness. It is also the first quality of a presenting scientist. There are many things to prepare. In fact, the host will often make a checklist prior to the event to make sure nothing is missing on the big day. There are just as many things on the checklist of the presenting scientist. Let us compare checklists.

The location: your lawn, next to your swimming pool, the public park barbecue pit…
The location: The meeting room, hotel ballroom, boardroom, etc.

Your high quality, appetising, fresh or defrosted food
Digestible, attractive, and novel slide contents

Enough skewers with marinated meat and brochettes of food for the guests, and some in reserve just in case they are very hungry
Enough slides for the time you have to present and some only to be shown if needed during the question and answer session

Your barbecue pit, cleaned, oiled, and with enough charcoal, and lighter fuel
The computer in the meeting room. You checked that it had all you required to present (software, fonts, players, codecs…)

Knowledge on how to operate the barbecue without having to call the fire engine for extinguishing your brush fire or the ambulance for treating your third degree burns
Knowledge on how to use the computer, microphone, presentation remote, and light control equipment without needing to be rescued by the technical staff

Alternative plans in case of rain (board games instead of ball games)
Your own PC, USB drive, paper handouts, shorter presentation

Your welcoming words and smile while looking directly at your guests as they arrive
Your welcoming words and smile while keeping eye contact with the audience

Prepared, organised, good-looking, knowledgeable, friendly
Prepared and rehearsed, organised, good-looking, knowledgeable, friendly

 The host is familiar with the hosting location: home. The scientist host is rarely familiar with the hosting location. Indeed, sometimes it is only known just before the beginning of the first session on the first day of the conference. That is also the time when you meet the chairperson, and discover the room and the computer used for presentations. You have much to do then. You have to check how to control the audio and lighting equipment, tell the chairperson how to pronounce your name correctly, check the operating system and the software (PowerPoint version and video codecs — for example QuickTime for Windows), install your presentation on that computer if required, etc. One cannot imagine a good host not being intimately acquainted with the hosting location!

 After a last check to see whether all is in order prior to the arrival of the guests, the host checks his or her appearance one last time, and then starts to relax, confident that everything is well prepared. The host is not under any stress while eagerly awaiting the guests. Their arrival fills the host with pleasure. These people have accepted the invitation and are honouring his house with their presence. They had a choice to do something else, but they instead decided to come. The happiness of the host is visible. He or she smiles while greeting them and thanking them for coming, and so do you. While smiling, the host looks at them (not sideways or above their heads) and greets them, conveying the pleasure their presence brings. The host remains attentive throughout the time shared with the guests. Do they need more food, more drinks? Do they have special requests? Are they smiling and happy?

 Similarly, you remain attentive during the whole presentation. Is the audience interested? Are you clear? Are you going too fast? Frequent eye contact with the audience provides the required answers. People may be noisy, shuffling uncomfortably in their seats, distracted, frowning, or on the contrary they may be still, returning eye contact, smiling, silent, looking interested. Taking the cues from the audience, you adjust contents and pace to maintain interest. A good scientist host builds in a safety time buffer for situations that require adjustment during the talk. If you need to slow down the pace and summarise what has been presented (with the screen behind you blanked), do not let that be at the expense of having to rush through the rest of your presentation. Keep time available for occasions such as this.

When the time comes for the host to interact directly with the guests, the attentive host does not let one guest monopolise all the attention and keeps watch to see whether other guests are discretely waiting to have a word. Similarly, during the question and answer session, you do not let anyone monopolise your attention at the expense of others. As host, enlist the help of the chairperson to deal with situations that reduce the effectiveness of a Q&A session.

The Visible Host (and the Co-Host)

Sleep little scientist, don't you pry

Vladimir zipped past his title slide to his second slide, the classic "outline" slide. He had seen it in all the previous presentations that morning apart from one (the one where the audience had clapped at the end). He turned his back to the audience and proceeded to read every single bulleted line in the outline, after which he turned back to his computer screen, clicked the mouse, and brought a text heavy objective slide, which seemed to be a downsized version of his abstract. He mostly read that too, this time looking at his computer screen. Somewhere at the back of the room, the door, which had been left ajar to let late participants enter, silently closed after two people had left the room. He was now in semi-darkness. The beam of hope, which inhabits any attendee at the start of a presentation, was rapidly dimming: this was going to be the usual presentation. The beam turned into candlelight glow when Vladimir brought the next slide entitled *"Methodology"*. The complex black and white diagram covered with font size 11 serif text had one attendee remove his glasses to wipe them on his shirt. Pasted from a PDF file original, the diagram suffered from the blurring that accompanies a combination of anti-aliasing technology and misaligned low resolution VGA projection. The blurring and Vladimir's lifeless tone unfocussed the attention of people as he pointed here and there to places on his computer screen to explain the various intricate algorithms used in the simulation model. The explanation lasted and lasted, to no avail. Cruising at high altitude, Vladimir had left his audience behind. Had he listened, he would have heard the noise of feet shuffling and seats creaking as people rested heavily against the backs of their chairs in preparation for a nap. The chairperson, sitting upright in front of the reclining people, looked like a watchtower. He had managed to keep one eye half opened for a few minutes, but resisting was futile. The volume of Vladimir's static voice hypnotically oscillated as he swung his head back and forth from the mike to the screen with the regularity of a pendulum. The audience was now comatose. Unknown to Vladimir, he had not just lost people's attention; he had lost all chances to get questions at the end of his talk.

Little scientists don't pry when they are sung a lullaby.

In the old times, presenter hosts were presenting alone in front of an audience. Today, they share the stage with a co-host: the computer. That co-host is so attractive, colourful, and animated that most of the people spend their time looking at the co-host, not at the host. Facing an information-rich screen, the brightness of which is enhanced by the dimmed lights in the room, the co-host reduces the role of the host to that of a narrator, a voice-over as they say in film production. The presenter, reduced to a disembodied voice, is now downgraded to the function of co-host whilst the co-host takes front stage as main host. Let not that be your fate. You deserve better than the computer co-host: YOU ARE THE HOST.

To retain your host qualities, you need to keep the co-host under control and here is how.

1) Remain in the limelight. Keep all the lights on. Be in charge of controlling the light level. If a slide requires dim lighting to be readable, ask the technician to bring down the lights and to bring them up with the next

slide. That way you can keep eye contact with your audience without being blinded by any light.

2) Whenever you can, use the "B" key, or blank the screen with a black slide; then engage the audience face to face. Deprived of its co-host, the audience only has you, the host, to look at! Do not rely on the slides to tell your whole story. Keep some parts for your one-on-one with the audience.

3) Be animated. Move. Use gestures. Reveal the dynamic person in you. There is nothing like movement to recapture audience attention. For your gestures to work, however, you must be seen, hence keep the lights on.

4) Keep needless gif file animations out of your screen so as not to distract your audience. Animations are great, but only if necessary to your purpose.

5) Emphasise your words using tones and gestures that re-centre the attention on you. Reveal the communicator in you. Use that disturbing silence that brings the audience back to you as it wonders why you stopped talking.

6) Simplify and trim down what is on your slides so that the audience absorbs their contents quickly and clearly. Do not let the audience linger on your slides because they are difficult to read, or because there is so much in them that it takes all their attention (leaving you, so to say, "outside of the picture").

7) Face the audience. Keep eye contact, as direct, honest, and authoritative people do, and as the direct, honest, and authoritative person you are. Do not turn your back on your audience. Use PowerPoint's custom animations or Keynote's build inspector to build layers of explanations and emphasis so that you do not have to turn and move a distracting laser beam across the screen.

8) Do not let the co-host say thank you! You are a responsible and polite host. Thank the audience yourself!

9) Do not fight with your co-host. When you bring a new slide, let the audience absorb its title. If you have prepared your transition, this should be fast and easy. If the audience does not know what to expect next, it will figure it out independently, and in the process only pay marginal attention to you. The co-host wins; you lose. Therefore, always prepare

your audience for what comes next. After a brief attention-focusing silence allowing the audience to read the title of your slide, guide the audience along a carefully prepared path through your slide contents.

10) Let the co-host help you if you have a language problem (heavy accent or stammering). In that situation, say less and show more, particularly in visuals.

Listen to the "Presenter's mistakes" podcast on the DVD

Being visible has its advantages; chief among them is a more subdued role for the computer co-host. It also allows the audience to put a name to a face, and a face to a scientific contribution. Being visible is not just a matter of lighting, or a matter of seeing your physical body, as the points above make clear. The audience also needs to discover your personality. If someone becomes a different person when presenting (and I have heard that comment many times — often in admiration), it means that that person's personality was exchanged temporarily with that of another role-playing person — the professional. Is that wrong? If that serious "professional" look contributed to project a positive, and authoritative image, do make sure that the Q&A session reinforces that perception — otherwise you will be pegged as a "fake", a superficial person. Also, use the Q&A session to reveal more traits of your personality, such as humour, honesty, kindness, or respect for others. In the end, these soft qualities influence the way your audience perceives your work as much as your hard scientific arguments.

Being visible from head to toe also has its potential problems. When you are behind the lectern, only the upper part of you is seen moving. Outside of that zone, everybody sees the whole of you moving. And that is exactly the place where you are reluctant to go for many reasons:

1) You are self-conscious about the way you look. You find yourself too fat, too tall, too ugly, or too skinny. You think people perceive you the way you

perceive yourself and feel that the less people see of you, the better. My advice is to forget about your self-image, because people have not come to a beauty contest, they have come to learn about you the scientist and your science. They see you as healthy, above average height, with strong facial features, lightweight, not as fat, tall, ugly, or skinny.

2) You are self-conscious of your body. Usually, you rarely stop life to look at yourself (men do this less often than women), but when you are standing on an elevated platform, or when you are caught by a spotlight, you become self-conscious of everything about you: the way you sound, the way you speak, the way you move, the way you stand. Your arms become real, your nose tip comes into view (it never did before), your empty hands seem to dangle aimlessly, your knees lock your legs straight, your immobile body starts feeling the pull of gravity, and your feet start complaining. Your new shoes bring themselves to your attention by squeaking and pinching your big toe. Your new tight skirt or leather belt that gives you the waist of a wasp is winning the battle against your rebelling abdominal muscles. At the same time, because your adrenaline is trying to turn you into a live wire presenter, your own physiological functions, normally a background noise, come into the foreground: strong heartbeat, fast breathing, sweaty palms, stressed bladder, cold hands, and shivering muscles. To find release, some unconsciously move the body into action. They move back and forth in a "bear dancing" pattern, resting on one foot and then the other, gently oscillating like a reed in the breeze. Others play with their footwear (high heel shoes), their presentation remote, their glasses, or with a marker pen that they cap and uncap 30 times a minute. Others hide their awkward hands in their large pockets (men) or their thumbs in tiny jean pockets, cowboy style (women). Some look away, or look down, or look at the screen... hoping that the audience will do likewise and not pay attention to them.

Effective release, however, comes from a good body position and comfortable (used) clothing. With your feet shoulder-width apart, feel the ground (you cannot do that with air-cushioned shoes, or soft rubber soles). Unlock your knees, as if ready to move. Take a deep slow breath. And then forget about your body to concentrate on your message. Let the message activate your body and do not let your body deactivate your message.

To enhance visibility, moving is good. It attracts attention. However, moving should have a purpose. No idle movement. No scratching. No rubbing. No distracting movement. No playing with objects — pens, glasses, laser pointers, remotes. Moving should be highly visible. Therefore, if you move your legs, walk at least one metre (several steps). If you move your arms, move the whole arm, as preachers or lawyers did in the past so that their words could be amplified by their broad sleeves flapping in the air like bird's wings (the French call it "*effets de manche*"). Do not restrain your arm movements to the lower arm while the upper arm remains stuck to your side. It shows you are tense, and it makes you look caged in and lacking confidence. There is a certain amount of exaggeration in the way you communicate on stage, compared to the way you communicate in a conversation. Your voice has to be louder so that you can be heard at a greater distance and your movements need to be amplified so that people sitting far away can see them. If you use large movements in a conversation at close range, you may hurt somebody!

Your clothing also attracts attention. However, if what it enhances visibly distracts, it is counter-productive. The colour of your clothes carries a message, but the message varies because different cultures interpret colours differently. It is therefore safer to dress in pastel colours, or black and white clothes, rather than to make a fashion statement with saffron, scarlet, or indigo coloured chiffon. Different countries with different climates also have different dress codes. Europeans tend to dress more formally and wear a suit and tie. People from California dress more casually than people from the East Coast do. Different professions follow different dress codes. People from engineering sciences tend to dress more casually than people from life sciences. Follow the smart casual dress code adopted by most people in your profession, and take into account the climate of the area where the conference is taking place. Remember also to wear clothes that allow the use of clip-on wireless mikes, and get rid of your uniform and your badge holder when you present. Your research centre jacket looks dull on stage, and the badge holder never won an award in beauty contests. Put them aside during your presentation.

Everything in and around your face, from its ornaments to the moustache, from the make-up to the colour of your teeth, has an effect on the audience. Your hairstyle attracts attention. Its colour makes a statement. It neatness makes a statement. Its length makes a statement. And its cut makes a statement. Your glasses attract attention. Their style makes a statement. Your earring(s) and tie make a statement. Observe yourself and determine what it is you want to convey to the audience through the way your face looks, and make sure that your appearance helps you achieve your goals.

Dressing appropriately is therefore not just a matter of personal taste. You should also respect the audience and avoid anything that might be seen as offensive or deviant. Instead of creating a difference through your physical appearance, your baseball cap or your gadgetry, consider creating that difference by the clarity of your slides and by your "stage presence". Presenters who are clear, friendly, respectful, and professional hosts are different.

The Hospitable Scientist

A speaker hosting a scientific presentation knows and prepares what the guests come for. If they do not know you, they come for knowledge. If they already know you, they may also come because they know your presentations are well prepared and interesting. They remember you from the last paper you presented at a previous conference. Your presentation created a benchmark. They expect a repeat performance. Consistency in quality presentations is the hallmark of great presenters. Your guests may even come because others told them you are a good presenter. Through word of mouth, your reputation grows. You are known as a talented scientist and an approachable speaker. You have a pleasant personality and great listening skills. Your willingness to share your knowledge with others is palpable. Your honesty, fairness, and respect for the work of other scientists are indisputable. You may be Italian, Russian, French, Chinese, American, Korean, Brazilian, or New Zealander... your accent or your country no longer matter. Even your affiliation to a prestigious research centre no longer plays a major role in the way people perceive you. You are you: a caring host, and a caring scientist, someone with scientific skills, interpersonal skills, and communication skills.

The world of science is competitive. Unscrupulous scientists, people you may even consider friends, may plagiarise, steal, and make false claims to establish antecedence in discovery. Publish or perish is all too real a saying. To survive or to remain in the same league class-conscious scientists may break ethic rules. Therefore, knowledge sharing needs wisdom as a companion. Sharing or not sharing (public domain or protected intellectual property) is a choice you rarely get to make because it is dictated by your employer. By publishing, you are putting your knowledge in the public domain. When you present to a public audience, even an audience of one, you are putting your knowledge in the public domain. There are rooms that a host does not open to unknown guests, rooms with valuable objects or untidy rooms. Similarly, during a public presentation, it would be foolish to disclose intellectual property that is not yet protected, or research results that could be the object of your next paper. You would not serve your guests half-cooked food. Similarly, it is not appropriate to disclose early results that are part of your ongoing research before thorough validation. You would not (some do) claim credit for preparing a dish that other people have made or that you bought at the store. And if you do use catered food and get compliments, you would mention the name of the caterer. Similarly, you would display intellectual honesty by disclosing the sources of the information displayed on your slides. This disclosure is vital to how other scientists perceive you.

You know that science is not done in a vacuum or in an ivory tower. Breakthroughs often occur when working in teams at the frontier between domains. Therefore, the presenting scientist often has an ulterior motive: that of finding research collaborators or partners. The host who throws an open invitation to a party often discovers several guests of interest. Such a host would be foolish to part from them at the close of the party without at least exchanging business cards to further develop a potentially fruitful relationship. The host scientist can only identify the people who have a genuine interest in his field through the questions they ask during or after the Q&A. Therefore, to benefit from a full or extended Q&A time, do not exceed your allotted presentation time. It is essential that information on how to reach you is made easily available to all attending the presentation (handouts, business cards, email address on slide, etc). It is also essential that you remain available outside the meeting room at the end of the talk so that interested people can approach you. Do look around you before going too deep into a lengthy dialogue with one person: others may be waiting on the side for a chance to talk to you. You may want to start distributing your business card first, before engaging in conversation. Some may be in a hurry to go to the next presentation.

A presentation is only successful if the Q&A is successful because the Q&A leaves the audience with its last impression of you. Hosting qualities are essential to a successful Q&A. The qualities respected in hosts worldwide are similar: they show respect toward their guests; they do not start an argument that discredits a guest; they are humble servants, attentive to the remarks or requests of their guests; they do not ignore what seems like minor requests; and they do their utmost to satisfy someone's requests without giving the other guests the impression that they are ignored. The same qualities apply to the presenting scientist. Whatever the question or the tone of the question, the scientist host respects the person who has had the courage to express doubts, lack of knowledge, or personal interest through a question. It is always better to be asked one question than to be asked none. Demonstrating respect toward other people's feelings and views is indeed possible so long as you continue to display an unshakable confidence in your own accomplishments. Remarks, or constructive comments from your audience, may be the way to personal scientific breakthroughs. Thank the person who offers the comments and do acknowledge your sincere interest in exploring the matter later. Give priority to no one. Any member of the audience is worthy of the same respect, regardless of social status or academic level. According to that principle, answering questions comes on a first come, first served basis*, and no question is worthless or regarded as such. You are relaxed, confident, never anxious, and always pleasant toward the people who accepted the invitation to attend your talk.

---

*See the entry "Who you're gonna ask" on the author's blog "*When the Scientist Presents*", at **www.scientific-presentations.com**

9

THE GRABBING VOICE

Speak with Confidence

Confidence is one of the principal ways the audience evaluates your expertise and determines whether or not what you say is believable. Naturally, you know that confidence comes from having rehearsed and being well prepared. You also know that confidence comes from seeing yourself as a host. However, despite all this, you may still be experiencing fear. Generally, so long as it does not last beyond the first 20 to 30 seconds of your talk, fear is considered with compassion by people in the audience. They recognise the signs: unstable voice (shaking), nervous cough, throat clearing, difficulty in swallowing or breathing, shaky hands betrayed by the trembling red dot of the laser beam on the white screen, no eye contact, closed postures (arms folded across the chest or hands clutched on the lectern), awareness of one's body and in particular these cumbersome dangling arms of ours, lifted tense shoulders, rapid speech, unsmiling face, and many more signs. So here are some techniques to regain that confidence and keep it throughout your talk (it's easy to lose it!).

Listen to the "David Peeble's argument" podcast on the DVD

Control Your Body

Control your adrenaline. You are the one creating your own fear, and you are the one who can get rid of it! In response to self-generated anxiety, your body, more specifically your adrenal gland, produces adrenaline. The effect of adrenaline is to prepare you to fight or to run away. Naturally, running away is not an option; therefore, it prepares you for a fight. Unfortunately, caffeine also helps produce adrenaline. If you are the anxious type, drinking tea or coffee before a talk is not such a great idea. Adrenaline diverts the blood from your skin (that explains the cold hands). The diverted blood goes to other organs, and in particular, your brain. That's the good news because you need lots of energy to think clearly. The other good news is that adrenaline has a short life. It does not remain in your blood forever because it is rapidly degraded by your own enzymes. Your symptoms of fear will disappear. Therefore, as soon as you start your talk, use your will power to focus on the task and stop paying attention to your fear symptoms because being aware that you are fearful makes you fearful (self-fulfilling prophecy), and adrenaline continues to be poured into your blood stream!

Control your breathing. Some people say, take a deep breath to calm down. Actually, if you are fearful, the adrenaline speeds up your breathing to add more oxygen into your blood stream. If you take a few rapid deep breaths, you only contribute to increasing the amount of oxygen, and may even feel light-headed because you are hyperventilating! Therefore, take a few slow breaths through the nose, and breathe out slowly through the mouth at a rate of one breath every six seconds or so. This should decrease your anxiety. You can even do that without anyone noticing as long as these are abdominal breaths. Some of you may say *I know what my abdomen is; it is my stomach. But what is an abdominal breath? I do not breathe through my stomach. When I breathe deeply, my upper chest expands and my shoulders rise.* If this is your case, you are not breathing properly, and your voice may lack confidence because it lacks volume and strength. With normal abdominal breathing, the shoulders do not move up and your chest does not move forward, but your stomach does. If you want to observe normal abdominal breathing, just lie on the floor and breathe deeply. Because your shoulders are pinned to the ground, they will not move and only your stomach will heave up and down as you breathe deeply. You need to breathe this way while standing up.

Control your muscular stress. Notice how high your shoulders are when you are stressed, so high that they seem to imprison your neck and head behind rigid muscle bars. Release your shoulders. Let them fall down and free your neck. Rotate your head a few times until you feel it moves around freely (do not do this in front of your audience, of course). Check your shoulders again. It is possible that they have gone back up already. Release them again. It is now time to take a final check to see if there is any residual stress that may affect the quality of your voice: your jaw muscles. Are they tight? Are your teeth clenched? Release them and sense how freely your tongue moves in your mouth. Finally, look straight in front of you and swallow some of your saliva. If you find it difficult to swallow, you are still too tense; if it is easy, you are ready to speak in a confident voice.

Control your hydration. Water is necessary for your vocal chords to be moist and your voice to be clear. Drinking room temperature water is recommended both prior to your talk and during a long talk. If your presentation lasts 45 minutes to one hour, the water you drank will slowly but most certainly make its way down to your bladder, and your bladder may (during your talk) signal its need to be discharged, particularly if it has not been emptied for quite some time. To avoid such disruptive "nature calls", empty your bladder prior to your talk; otherwise, rushing through your conclusions and Q&A to respond to your pressing need, and disappearing right after your talk when people are looking for you with more questions, may be misinterpreted.

Control Your Mind

Rehearse and memorise the first minute of your talk. Doing this enables you to relax more easily. You are in "auto-pilot" mode. The memory is in charge of what to say; your brain is released from that duty, at least for the next minute of your talk. You can spend the excess brain cycles to concentrate on how to say what you have to say in a smiling, friendly, and engaging manner. In this way, your fear is not even seen. By the time the adrenaline rush is over, you will have successfully overcome the hurdle of the presentation cold start, and displayed great confidence in your speech.

Calibrate your expression on a reference standard. Talma was Napoleon Bonaparte's favourite actor. In his novel, *les Misérables*, Victor Hugo mentions them as being seen "arm in arm". Talma had found a way to always keep the same happy and joyful expression when appearing in front of his audience. While still backstage where many people were actively working, immediately before entering the stage, he would approach someone and ask, "*Dear Sir, would you kindly tell me what time it is?*" After being told, he would then kindly smile and politely say, "*Thank you Sir*", and on that note, keeping the same smiling expression on his face, he would immediately walk on stage and speak his first words using the same tone of voice. His smile was never fake, always genuine. So right before you start your presentation, imagine yourself very happy to see that these people have accepted your invitation, and smile as you say " *Good [morning, afternoon, evening] ladies and gentlemen, fellow scientists. Thank you. Thank you for being here, and for your interest in this topic.*" Then watch the positive response of your audience at the sound of your warm greeting and the sight of your great confidence.

Be positively charged by the certainty of the worth of your contribution. Your conviction is carried by your confidence. The audience may disbelieve the facts you are presenting, but they will have not doubt that you believe in them. How can you convince people of anything if they sense your uncertainty! People hear more than words. They hear happiness in the way you say your words. They hear your smile even without seeing your face. They hear your boredom even though your words may seem enthusiastic. If the way you sound is not aligned with your words, they hear the gap and disbelief sets in. If your gestures are not aligned with your words, they see the gap. For example, when some people see you touching or rubbing your nose when you speak about your results, they may interpret that gesture as possible doubt, or lack of assurance.

Be positively charged, particularly at these times in your talk where lasting impressions are formed: your hook slide, and your conclusion slide. In my experience, scientists fail in their paper, at the same place as they fail in their presentation: the conclusion. It is often a lacklustre conclusion, an unavoidable formality one has to go through at the end of a talk. The conclusion carries no dynamism, no conviction, nothing to convey to the audience the significance of the scientific contribution to the audience, be it potential or real. The end of the presentation is flat, as flat as the energy level of the presenter who, on the finishing line, ends up reading the points on the last slide as one reads from a shopping list. Rehearse your conclusion as much as you rehearse your introduction. Keep it positively charged. Carry your confidence forward to the very end of your talk.

Control Your Eyes

Target a member in the audience who is obviously paying attention to you. If you are uncomfortable in the first few seconds you are facing your audience, look at someone who seems responsive and maintains eye contact with you. This will shift you from the situation of talking to a seemingly impersonal, intimidating crowd to that of addressing yourself to someone, as in a conversation over the fence with a neighbour. Conversation is a much more familiar mode of communication, and without much of the public speaking trauma. Naturally, you cannot continue doing so for the rest of your talk, because the rest of the audience will feel left out and the individual you are focusing on will feel uncomfortable and wonder why he or she is worthy of so much of your attention. Because you are talking to someone who is far away from you, raise your voice to be heard. Do not adopt the typical one-on-one conversation tone; it is not loud enough, even with a microphone clipped to the lapel of your shirt or blouse. Targeting someone enables you to keep your composure and your confidence.

Do not frequently shift eye contact or scan the audience. Scanning is impersonal. If you start a sentence while looking in a direction, finish that sentence looking in the same direction before making eye contact with another part of the audience. In this manner, people feel you are speaking to them individually. People take longer to be convinced if they are deprived of eye contact. To be able to keep eye contact with your audience as long as you possibly can, you need to be so familiar with the contents and sequence of your slides that you no longer need to look at them to remember what is on them.

Control Your Clothes

Wear clean clothes. I remember having to talk after a spaghetti lunch once. Tomato sauce spotted my white shirt right where everybody could see it. I tried but could not completely wash out the stain, and I did not have access to another shirt in such a short time. As a result, I could not focus totally on my talk. What were people thinking about a guy speaking with a great big pinkish stain on his shirt? I definitely lacked confidence, stood behind the lectern as much as I could, held the microphone high up at an unnatural angle trying to cover the spot, and frequently faced the screen. To speak confidently, you must feel good about the way you look.

Wear comfortable used clothes. New clothes, especially new shoes, new shirt, new trousers or skirt that make you look good are often uncomfortably tight or uncomfortably short once standing on a tall platform. They take your mind off your main purpose: presenting. Furthermore, a tight shirt, jeans, or skirt may compress the diaphragm to the point that your breathing (hence the quality of your voice) is affected.

Control Your Lips

Do not belittle yourself. People say the strangest things when in front of other people: they spontaneously draw attention to their shortcomings! I remember this short presenter, whose first words when he came behind the lectern were, *I'm sorry. I cannot see you because I am short*. He was short and had to lower the lectern microphone all the way down. The audience could not see much of him behind the lectern but could hear his voice fine. This person knows he is short. Therefore, before a presentation, he should go to the chair and ask to have a small platform installed behind the lectern to stand on and be seen. Alternatively, he could ask for a wireless microphone and a presentation remote, so he could move away from the lectern altogether. There is no need to be sorry for being who you are. Be proactive and find ways to reduce whatever threatens your effectiveness as a presenter.

Do not say that you are nervous, or that you hate speaking in public at the beginning of your talk. These remarks are suicidal. The audience does not respect people who complain about themselves. When you present to an audience, something strange happens: you are becoming the message, and the message is becoming you. When you say bad things about you, the audience stops trusting you AND your message. You lose credibility extremely fast this way.

Speak for Intelligibility

"Intelligible" is an adjective, and therefore subjective. English spoken with a very strong French accent may be intelligible to the other French people in the audience, but not to the Chinese. However, a Chinese scientist whose colleague is French may have become quite familiar with the French accent. To complicate things, each country has its regional accents, or accents based on belonging to a certain class in society. This accent also colours the way a non-native English scientist speaks. Beyond accent, speech can be intelligible to one but not to another if the words used are understood by one, but not by the other. Mezzanine financing is a term known to a venture capitalist, but not to an engineer. Knowledge is essential to intelligibility. Lack of familiarity with accent and jargon are just two of the factors that may reduce speech intelligibility. There are others: fast speech, a soft quiet voice, a noisy microphone or no microphone... They all affect our ability to make up intelligible words out of sounds.

 Listen to the "Dealing with accent" podcast on the DVD

To enhance intelligibility, the presenter is not as helpless as it appears. Some remedial actions bear fruit over the long-term, others are short-term palliative measures, and others still are actions with an immediate noticeable improvement.

Actions With an Immediate Noticeable Improvement

Speak more slowly (but not too slowly) and pause where you would use punctuation in writing. Did you notice that Heads of States usually speak slowly, not more than 120 words a minute, and sometimes less? Did you notice how they pause after they say something of importance? This slow pacing gives them time to think and time to study the reactions of the audience. It also gives the audience time to understand, and journalists time to write down what they hear without misquoting!

In scientific presentations, speaking slowly is even more necessary because many in the audience are probably not native English speakers. Recall the time when you were learning a language. Recall how difficult it was and how fast your teachers seemed to speak when in fact they spoke at normal speed.

Some people do speak fast. They have always spoken fast. To them, slowing down is nearly impossible. They take presentation skills classes one after another to be regularly told that they speak too fast. Can they speak more slowly? They can and they do. For example, when they speak to an elderly person who is hard-of-hearing or to a toddler, they naturally slow down. Usually, as they speak more slowly, they speak louder. The two seem to come together, as if intelligibility was enhanced by speaking loudly, when it fact, it is mostly enhanced by clearly articulating the syllables in a word and the words in a sentence. When Mr. Rogers, a famous presenter of children programs on American TV, spoke, he did not raise his voice — he paced his speech. It is as if he pronounced words at the speed of his heartbeat (I am not advocating such regularity in speech, however). Therefore, when you speak in front of an audience, pretend to yourself that most of the people in the room are foreigners unfamiliar with your accent and with your jargon — they probably are.

Even if you are not a speaker who naturally speaks fast, nervousness (and the fear of exceeding your presentation time given your many slides) causes you to speak faster; therefore, reduce the number of your slides to fit a more gentle presentation pace. Since keeping your nerves under control is essential, a few deep breaths (inaudible ones) before starting will help, and so will refraining from drinking caffeine-laden beverages prior to presenting.

Some people naturally speak too slowly. They may be confident lecturers, and indeed talented scientists. Yet, they speak as if their audience is made of sophomore students when people are well past that stage. Audience frustration is increased when the presenter repeats or clarifies (slowly) what the screen already shows clearly, ignoring that the speed of reading is at least twice that of listening.

What are the benefits of slower speech? It forces you to apply the principle *less is more*. Since you must say less in the same amount of time, you will inevitably need to reduce the material you intend to present, or improve your visuals so that they become simpler and no longer require extensive explanation. Furthermore, since you are more intelligible, the audience enjoys you and your presentation more.

Stand correctly to be heard correctly. An incorrect posture impedes the quality of your speech. The correct standing position is NOT the chin-up chest-out position a soldier adopts when saluting a superior. That position restrains the expansion of your diaphragm when breathing in. The natural position is one where the body stands firmly connected to the ground, feeling the ground but not clutching it, not bent backward resting on the heels, and not bent forward resting on the balls of the feet, but resting firmly and equally on both. The knees are not locked. The legs are not joined but are slightly apart so that your feet are aligned with your shoulders. The shoulders should not be moved up by tenseness, or slumped over a hunched back, but squarely resting ever so slightly back on an upright spine. Your shoulders are the clothes hanger supporting your arms. The arms are not folded in front of the body compressing the diaphragm, or folded behind the body forcing the shoulders back. The arms drop loosely along the side. The head faces the audience, straight (not at an angle) as would the head of a puppet pulled up by strings, hanging loose on the shoulders, able to move freely. In this position, the air from your lungs freely flows and your voice is strong and has presence without effort. You know your air passage is totally free when it is easy to swallow. Try standing correctly and swallow. You will be amazed how easy it is. After this, stand in different positions and try to swallow. Notice how much your posture affects the passage of air.

The body betrays its fear of public speaking by tensing throat and jaws, and by lifting shoulders and upper chest, thus impacting the free flow of air in and out of your lungs. I have even seen people so frightened that they appear short of breath and keep swallowing with difficulty. Fear has an immediate effect on your breath, and therefore on your tone of voice, removing its normal assuredness and confidence. To get back to normal, release your shoulders. Let gravity take over. It is sometimes helpful to rotate the shoulders and head a few times to let them return to their normal position. If you are speaking and feel the tension returning, briefly lift your shoulders up and down to bring instant relief (do that discretely while the audience is watching the screen!). And do not forget to relax your jaws — your teeth should not touch each other. Focus on doing that. Get your body to relax, and taste the sweet sense of calm that precedes a clear audible voice.

Voice the end of your sentences. Hearing the end of a sentence is important. In English, new information about a topic is often expressed after the verb, thus giving more emphasis to what is said at the end of a sentence; therefore, speak up so that the audience does not have to guess what you say at the end of your sentences. In English, it is customary to start a sentence on a high pitch, and to finish a sentence on a lower pitch. It helps distinguish the start of a new sentence, or the expression of a new idea. However, a lower pitch is not equivalent to a lower volume of your voice. The volume of your voice is regulated by the volume of air passing through your vocal chords (the lesser the amount of air, the lesser the loudness). A lower pitch is simply a lower tone. When people run out of breath before the end of their (often long) sentence, they automatically reduce loudness and tone. If they do not, they usually strain their vocal chords. Whichever way you look at it, it is bad: bad for the audience who does not hear clearly the end of your sentences, or bad for your vocal chords. There must be enough air left to sustain your loud and clear voice until the end of your sentences.

The good news is that our ears have automatic gain control. If a voice is too soft, the outer hair cells located along the length of the coil of our cochlea act as mechanical amplifiers. Another good news is that we are good at guessing what we do not hear by cleverly using the context in which the words are said, or even using partially heard words, or reading lips to fill in the missing words. The bad news is that microphones do not have automatic gain control. When you move your mouth away from the microphone, the microphone does not compensate for the decrease in loudness. To compound the problem, the words used in a scientific presentation are not ordinary words and the context of these inaudible words may not be that helpful to guess what they are. Therefore, keep your voice loudness constant at the end of a sentence. Do not mumble, or mutter.

This is only possible if 1) the air you keep in your lungs is enough for even long sentences to be voiced without loss of loudness, 2) your posture does not prevent the air from getting out of your throat freely, 3) you pace your words, breathing during the gaps where punctuation would have been in a written sentence, and 4) you keep the distance from your mouth to the mike as constant as possible, unless you intend to raise or decrease loudness to attract attention.

Long-Term Actions with Guaranteed Long-Term Ameliorative Effects

Identify the words you mispronounce and correct your pronunciation. In your field, the number of keywords is finite. Some of them appear multiple times in your paper. Identify these. Which ones are they? The first list contains the keywords in your paper. Usually, the author supplies such a list to the journal. A more extensive list could be derived from the scientific keywords in your abstract. They are highly representative of your work (if your abstract is well written). The abstract is short, and extracting a list of keywords should be fast. Your final option is to run a word frequency counter program over your whole paper or the text on your slides. There are a number of them available for a fee such as the advanced version of the Hermetic Word Frequency Counter for PC suitable for the long hyphenated names found in chemistry or life sciences or "Word Counter", a freeware program for the Mac. Focus on the long words with high frequency, as these are usually the most difficult to understand if mispronounced. Once you have your list of words, you have three choices: ask a native English speaker to check your pronunciation of these keywords, look up your words in a dictionary that has an online pronunciation of the word such as the one found here:

http://www.merriam-webster.com/dictionary/X
(replace the X with your word and click on the speaker icon)

Alternatively, you could also use a good text-to-speech program and compare the computer's pronunciation with your own pronunciation. Apple has a great text-to-speech program. It is part of the Mac OSX operating system. Access the system preferences and select speech. Then select the text-to-speech tab and select the "Alex" voice. You are now only one step away from hearing what words sound like. Select "Speak the text selected when the key is pressed" and click on the set key button. Choose your key combination. Typing this key makes the computer speak out the highlighted text.

It only takes a few mispronounced words to considerably reduce intelligibility. I recommend that you check your pronunciation if you are not a native English speaker. You may be amazed that some words you thought you pronounced well are actually mispronounced. Some of these mispronunciations may even be embarrassing. You may for example mispronounce the "i" diphthong, (the "i" of smile) so that it sounds like "smell". I am sure that your presentation skills tutor never asked you to "smell" while facing the audience.

Improve your speech fluency by talking to English-speaking people. Scientists of a given nationality tend to stay together. They speak their native tongue in their labs, or read newspapers written in their native language.

They may have a spouse from their homeland and the language they speak at home is not English. When they speak English, the sentence construction is still greatly influenced by the grammar of their native language. As a result, they make very slow progress in improving the intelligibility of their English. Their speech is hesitant; words come out in bursts followed by pauses in the wrong places as they translate their thoughts into English. To speak English fluently, there is no magic wand. You need to place yourself in an English-speaking environment. By constant practice of the spoken language, your speech will become more continuous. Your objective is to be intelligible to an international audience, and that has a lot to do with the regularity of your speech and with not translating thoughts, instead thinking them directly in English (which cannot be done if your vocabulary is limited).

Understand what your country's accent is, and decrease it. Your objective is not to lose the native tang of an accent that gives flavour to your talk. Keep that charming accent of yours, but remove the parts that make your speech unintelligible. To that end, you can do three things: First, repeat what you hear, not just the words but also the intonation. Then, learn about your accent and how it differs from a standard British or American accent. For that, visit Wikipedia

(http://en.wikipedia.org/wiki/Non-native_pronunciations_of_English) and (http://en.wikipedia.org/wiki/Accent_reduction)

Finally, work on reducing your accent.

I remember learning the English pronunciation as a young man through phonetics and an English dictionary. If you can recognise the English phonemes and learn to pronounce them accurately, then with the help of a dictionary, you can learn how to pronounce any English word. To learn the English phonemes, check out
http://en.wikipedia.org/wiki/English_phonemes.

To hear how they are pronounced, download a free program from the British Council and listen to them on your computer.
http://www.teaching-english.org.uk/try/resources/pronunciation/phonemic-chart

Alternatively, you could use a text-to-speech program (the Apple "Alex" voice is quite accurate) or have a native speaker say those for you.

Speak for Attention

Grab the Attention

Change the volume, pitch, and pace of your voice. We are very sensitive to change. Anything that changes is like an alert, an automatic, unreasoned call to attention. A voice change grabs attention. The change in volume often occurs at the beginning of the sentence expressing a new idea. This change is usually preceded by a pause that helps the listener identify the change. Punctuation (pauses for breathing) contributes to the volume of your voice since the lungs, refilled with air, have reserve air for greater volume. Another change is the change in pitch. The voice moves up or down one register. Sometimes that change is combined with a change in loudness. Sometimes it occurs on a phrase or word to attract the attention of the listener to it. The speaker can lower the pitch in anticipation of a later rise in the same sentence (or in the next sentence) to highlight a point of importance. And finally, there is the change in pace. The voice slows down to emphasise a short phrase by increasing the delay between its words, thus not voicing the liaisons between words, or by emphasising a word by articulating all the syllables in that word.

These three changes (volume, pitch, and pace) always take place while the speaker maintains eye contact with the audience. They are enhanced by the pause that precedes and follows them. The punctuating silence makes the change effective. Silence is itself a change from speaking. This change also grabs the attention. The listener expects continuation of the speech and instead finds an arresting, ear-perking, and eye-widening silence. Presenters use silence to get the audience to refocus on them.

Done without the intent to emphasise, these three changes are detrimental to the talk. Some people change pace because they are running out of presentation time. Some people mark a slight pause after an article pronounced on a higher pitch such as "a" or "the" as if to provoke the . . . [listener] to fill in the . . . [pause] with the . . . [delayed] word. Some people speak words in bursts and then either repeat them, pause, or use fillers such as "um". These changes tire the audience.

A change in loudness can also be detrimental to the talk, such as when the presenter turns the head away from the microphone or toward it, thus raising or lowering the volume for no purpose. Undue silence attracts the attention but it also distracts the listener (for example, the speaker looks for something in his/her notes, fiddles with the computer mouse, or drinks water).

Use words calling the attention. Some words naturally attract the attention even if the audience is not looking at the speaker. These words express a change in the course of the presentation such as a new slide or a new section: "Let us now examine ...", "On the upcoming slide...." or "In conclusion". Attention-grabbing words also express a change in the argument ("however", "but", "in contrast") or surprise ("curiously", "unexpectedly", "surprisingly", "interestingly", "strangely").

Questions. Questions are powerful immediate attention-grabbers. Use any question word (why, how, what, where, who, which) and watch how the audience reacts. If silence had left, it returns. If people had lost eye contact, they resume it. Furthermore, questions require a change in pitch and intonation. They end on a pitch different from the usual lower sentence-end pitch. If a question is asked, an answer is expected. The audience wants to know. Questions are great when used as transition devices to introduce a forthcoming slide (ask the question before you bring the slide that answers it, not afterward when the audience is looking at the screen and not really listening to the question).

Sustain the Attention

Slow down your speech. People will drift in and out of your speech. They may even be ahead of you, but most likely, they will lag behind trying to follow you. Therefore, to sustain attention, it is necessary never to leave your audience behind. Pace your speech so that people can follow, particularly if you have a strong accent. One hundred and twenty words per minute is adequate even though the natural speaking rate may be 165 to 180 words per minute.

Tell your contribution as a story. People are attentive to stories. In your paper, you rarely disclose the story of the research, how you decided on the method, how you had to change some initial assumptions once the data came in, how you were surprised by some results and what you did to try and explain them. If you have enough time to present, you can structure your presentation as a story that highlights your contribution. There is no need, however, to reveal the errors due to inappropriate, sloppy, or faulty research processes unless these errors led to a breakthrough. The tone of a story is natural, not formal or stern. It features you, the researcher, and it features your results. It is personal. It is much closer to what really happened during your research than what was written in your paper (not that you lied in your paper, but then you downplayed the challenges you faced). Telling a story is always more animated than merely presenting results.

Ask a question with a delayed answer. The power of the question has been stated already. Not all questions are equal. Some are answered immediately. Some are fundamental questions presented in the introduction of your talk and the rest of your talk endeavours to answer them. The audience is kept attentive simply because it wants these questions answered.

Speak for Persuasion

You may ask, what do people need to be persuaded of? They need to be persuaded of three things. 1) you are an excellent researcher, 2) your results can be trusted, and 3) people can benefit from your research.

Speak Out of Knowledge Without Hesitation

Have your rehearsed speech and gestures carry your conviction. Confidence is gained through many rehearsals. Rehearsals help you gain fluidity in speech and remove any trace of hesitation: the annoying "um" and "er" fillers are symptomatic of a brain trying to verbalise deeply buried information. Rehearsals bring to your tongue the words to express your thoughts. They set the tracks for your train of thoughts. Some in the audience interpret hesitation in speech as prudence, others as lack of knowledge or as a sign of unpreparedness. Why give your audience the opportunity to misread you! Rehearse to convince.

Speak out of knowledge. Vague words lessen your authority as an expert. Generalities fail to convince. They may even appear as an attempt to convince without evidence or on the wrong evidence. If "many" people are in your research field, it does not mean that your results are worthy of praise. Other researchers may use the same method as you, but that does not guarantee that this method is the best. Who these researchers are is as convincing as what they use. As an expert yourself, you know their names. You know the reason(s) why your method is better than other people's methods, why your data is better than other people's data. When you speak, the expertise naturally flows out of the precise words you use. You advance "educated" guesses or interpretations. Your suggestions are believable. Uneducated guesses or lack of explanation for what is presented on the slides reveal shallowness of research, and inevitably, shallowness of the researcher. The contents reflect the container.

When you rehearse your talk, identify the possible reasons for the overall shape of the curves on your graphs: their kinks, their trends, their limits, their differences — and explain them out of knowledge during your talk or during the Question and Answer session.

Establish Credibility Early

Present yourself in your research context. Belonging to an institute known for the quality of its research in your particular field speaks in your favour. Here again, the contents reflect the container. If your institute is not well known, but the principal investigator or senior researcher contributing to your paper is well known in your field of research, mention him or her by name.

Listen to the "Personal credibility" podcast on the DVD

Present other people's work, and present it in a positive light. It is so tempting to attempt to look good at other people's expense, or without them. Yet, without them, you probably would not even be presenting your paper! When comparing your work with theirs, be fair. Explain why you get better results, why your method is more suitable for given data, why the maths play in your favour when it comes to computational load. Claiming superiority without giving others due credit is viewed by some as inconsiderate. It makes them suspicious (were the results "cooked" to impress? Was the data source biased?) Presenting and answering a counter argument against your own argument during your talk pre-empts aggressive questions during the Question and Answer session, while also strengthening your own argument. People can more easily believe someone who is willing to consider a different point of view.

Emphasise in Various Ways

Repeat. Contrary to the written word, verbatim repetition of the spoken word is so usual in the way we speak that we do not consider such repetition out of place. Emphasis makes repetition worthwhile. Alas, repetition is often misused by speakers who repeat to avoid silence when they run out of words or ideas.

Frame the raising and lowering of your voice with silence. The brief punctuating silence surrounding the volume changes makes these all the more noticeable.

Reduce your pace, pause between words, articulate each word, and add movements as if you were hammering your words. In this way, every single word spoken receives emphasis. A famous example of this technique is the "read my lips: no new taxes" said by past US President George W. Bush Senior. (You'll find a low quality voice sample of this famous sentence here:

 http://en.wikipedia.org/wiki/Read_my_lips:_no_new_taxes

Another interesting example is the famous "*Je vous ai compris*" speech of General De Gaulle. Note his body language. You can find it here:
http://www.ina.fr/archivespourtous/index.php?vue=notice&id_notice=I00012428

Note that what was said in these two examples was short. If you have something important to say, keep it short.

Say that something is important. You can say it with silence. Remaining silent after a sentence automatically emphasises this sentence. People have the time to process it and to have its meaning "sink in". Besides silence, you could also use words to attract the attention of the reader to the importance of what you have said, or what you are about to say: *What I am about to say is really important, Here is the most significant aspect of my contribution, Pay attention to this, First and foremost, I would like to draw your attention to* . . .

10

THE ANSWERABLE SCIENTIST

A successful question and answer session erases the bad impression made by a mediocre presentation. Some presenters come alive during their Q&A. Their expertise is clearly seen in the way they handle difficult questions or questions requiring precise answers. They are obviously more at ease interacting and conversing one-on-one with people than going into a monologue for 15 minutes. Some presenters shine in both presentation and Q&A, making a strong impression on the audience. Alas, some presenters also collapse during an ill-prepared or hostile Q&A, even after presenting well. They fail because they lack skills (reasons 1 to 3) or they lack preparation (reasons 4 and 5).

Reason 1: Inability to perceive that the role played by the presenting scientist in front of an audience is that of a gracious, respectful, and attentive host
Reason 2: Skills insufficient to handle questions from a public audience
Reason 3: Lack of expected knowledge or expertise (from the point of view of the audience)
Reason 4: Judgmental slides, errors in logic, or plagiarism found on the slides
Reason 5: No effort to anticipate questions, in particular questions related to the slides

To prevent such failures, it is essential to be familiar with the process of answering questions from a public audience (reasons 1 and 2 above), to go through the slides during the rehearsals and anticipate possible questions (reasons 4 and 5), and finally to recognise the various traps set by questions and questioners, and learn trap avoidance techniques (reason 3).

Vladimir on the grill

Vladimir ended his talk with the usual "thank you for listening to my presentation". A large redundant "thank you" filled the screen behind him. Yet, even though both his words and the message expressed thanks, his posture and face did not. Arms folded across his chest, he was waiting for questions. The Director turned toward the board members and asked them if they had questions, secretly hoping they did not. The blood pressure of Vladimir's manager immediately increased, and started to decrease after a five-second silence when it looked as though there would be no questions. However, a board member who had been very quiet during the previous presentations slowly came out of hibernation. Professor Linda Sinclair, from Cornell University, looked left and right at her colleagues, and said, "well, if you gentlemen do not have questions, I have." And with that statement, she launched the first question she had jotted down on her pad during Vladimir's talk.

"Doctor Toldoff, could we please return to your conclusion slide."

Vladimir pressed the left arrow key on his presentation remote.

"I refer to point three on your conclusion slide. You have omitted to say on what basis you came to that somewhat far-fetched conclusion. Would you mind enlightening us? My second question . . .

Vladimir panicked when he heard "omitted" and panicked even more when he heard "far-fetched". She disagrees with my findings, he thought. His ears heard the rest of the second question — "how convinced are you of what is stated in point three" — but he was not listening. He was thinking about what to respond to her first question.

To Vladimir's manager and to the Director, this was a situation of clear and present danger. Before Vladimir could speak his manager got up to answer, but Professor Sinclair interrupted him with a hand gesture.

"Thank you, but I would like to hear the answer from the lips of the person who conducted the research. Dr. Toldoff, may we have your answer?"

The heat had been turned on, and Vladimir was on the grill. Some of his conclusions, and in particular, conclusion #3, were tentative, based on promising experiments that still required further work to move beyond reasonable doubt. Vladimir tried to recall where in his 25 slides were the ones he used to support conclusion #3. He could not escape from the questions, but PowerPoint let him escape from the slide show mode so that he could look at the miniature slides. Professor Sinclair waited patiently as Vladimir fumbled with the mouse and clicked here and there, searching for a supportive slide. Finally, he found slide 17, brought it to the screen, cleared his throat, and spoke.

"As I said before and as the data on this slide show..." Professor Sinclair stopped him in his tracks.

"Doctor Toldoff, are you implying that I have not been listening, and are you going to simply repeat what you have said before?"

Taken off balance by the remark, Vladimir paused and looked anxiously in the direction of his manager for help. None came, but the Director came to the rescue.

"Vladimir, why don't you rephrase the question and answer it as best you can."

Vladimir only remembered the "far-fetched" part of the question. He mumbled, "Professor Sinclair disagrees with point three of the conclusion and would like to know why I concluded this?"

Professor Sinclair corrected him.

"Doctor Toldoff as I have said before, and you should have listened, I do not disagree with your point #3, I merely ask to see additional supporting evidence, if any, and I also would like to have you share with the board your own level of confidence regarding that conclusion."

This time Vladimir had listened, but he knew he could not satisfactorily answer her question because he had not prepared extra slides with other supporting evidence, thinking it would not be necessary given the short time each had to present. He knew that conclusion #3 was tentative, but thought that if he admitted his lack of confidence, Professor Sinclair would start a long line of questioning that would lead to the total destruction of his work. In a meek voice, he uttered the sentence of his downfall.

"On the basis of this diagram, I am confident that this conclusion can be made." A foreboding silence, the type that precedes a storm, fell upon the room. Professor Sinclair glared at Vladimir. On her left and right, other board members became agitated. One, Dr. Zhang, became tense and broke the silence by quickly pressing and releasing the spring-actuated mechanism of his pen, producing a series of explosive clicks. The other, Professor Takayushi, was throwing rapid puffs of cigarette smoke into the air as volcanoes before an eruption. When he crushed his cigarette butt in the ashtray, she spoke.

"And I am confident that this conclusion cannot be made, Doctor Toldoff, and here is why. Firstly..."

Vladimir was no longer on the grill. He was in the fire.

Let us now see what would have happened if Dr. Sorpong, a man with great presentation skills, had been asked to present the same slides in front of the board.

Dr. Sorpong on the grill

Dr. Sorpong ended his talk with his usual smile accompanied by a warm "thank you for your kind attention." With the conclusion slide still on the screen, he then moved toward the board members, and in an inviting gesture, arms stretched in front of him, palms up and still smiling, he offered: "I would be delighted to respond to any question you might have." The Director turned expectedly toward the board members. A board member who had been very quiet during the previous presentations came out of her reserve. Professor Linda Sinclair, from Cornell University, looked left and right at her colleagues, and said, "well, if you gentlemen do not have questions, I have." And with that statement, she launched the first question she had jotted down on her pad during Dr. Sorpong's talk.

"Doctor Sorpong, thank you very much for this nice presentation. Would it be possible for you to develop point three of your conclusions, and in particular tell us if you have more substantive evidence to support it? My second question is related to the first: How confident are you of what you stated in point three?"

Dr. Sorpong had been carefully listening to both questions. He rephrased them to give himself time to think of a good answer.

"To justify our third claim, you feel that more than the simple diagram from slide (he briefly looked at the condensed print out version of his slides)... 17 is needed, and you would like to know my level of confidence in drawing out this conclusion." Dr. Sinclair nodded her approval. And with that, he typed 17, pressed the enter key, and slide 17 immediately appeared on the screen. Dr. Sorpong's manager looked with pride to his star researcher. The Director wondered how the correct slide appeared so fast and made a mental note to ask Dr. Sorpong how he did that.

"Well, Professor Sinclair, in answer to your first question, I would say that you correctly identified that our third conclusion is indeed tentative and based on promising experiments that still require further work to reach a higher level of certainty. In answer to your second question regarding my level of confidence in what is expressed in point three, I must say that I am partially satisfied at this time, but that I eagerly await the results of our ongoing research to bolster my confidence. Professor Sinclair, I have here a more complete and detailed version of the simplified diagram on this slide. Would you like to see it?"

"Yes please, I would like that."

As he moved the mouse, the arrow cursor appeared, but as it reached a corner of the screen, it changed to a finger. Dr. Sorpong clicked the mouse button and a full version of the diagram appeared. The Director seized his pen and scribbled "must congratulate Dr. Sorpong for his presentation skills" on his notepad.

The Process of Answering Questions

Question and answer sessions in scientific events are often facilitated by a chairperson. Besides handling the opening statements for the sessions, the chairperson plays a role the presenter cannot ignore. The chairperson is the timekeeper, and as such, sometimes signals the remaining presentation time. The chairperson is the facilitator introducing each presenter and presentation topic in the context of the overall theme of the sessions, thus bringing a sense of unity. The facilitator is also the one asking the first question when the audience is slow in asking. Finally, the chairperson is also known as the moderator. The verb *moderate* signifies *to make less intense*. There are times when the relation between the presenter and the audience is intense, as when a person (not just what is presented) is personally attacked. The moderator would then intervene to state the boundaries for a fair scientific exchange. The moderator also intervenes when the Q&A session is hijacked by members of the audience (rambling comments, audience-led discussions, continuous questions from one individual only).

Listen to "The chairperson and the presenter" podcast on the DVD

To answer questions from a scientific audience, it is best to follow a seven-step process characterised by the letters shown in Fig. 10.1.

What you present on your slides limits the range of possible questions from non-experts because they rarely ask questions on what has not been presented. Questions need to be prepared ahead of time (facing an audience while unaware of the types of questions people could ask is definitely too risky). You normally prepare the questions alone sitting in front of your screen and going through your slides, one by one, jotting down the questions that each slide could possibly raise (Fig. 10.2 suggests a format for the question preparation sheet). Formulate the question and put a check mark in one of the four boxes. You could remove what caused the question from the slide or you could modify your slide so that the question no longer comes. You could leave the slide unchanged and briefly answer the question orally. You could add an extra slide after the conclusion slide (the supplementary slide would only be

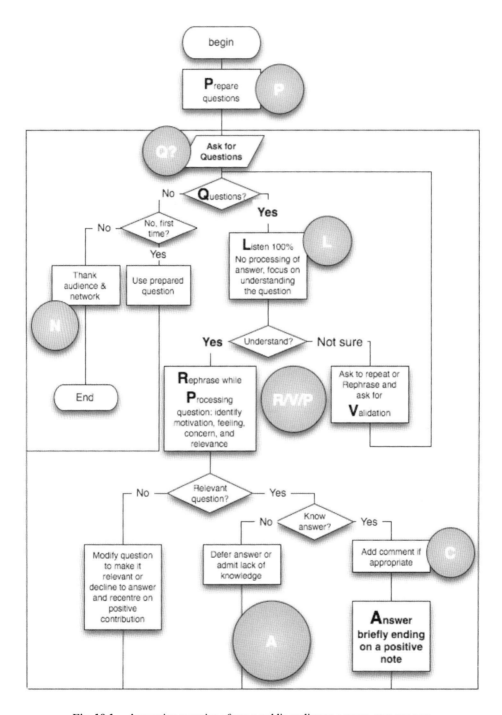

Fig. 10.1. Answering questions from a public audience: a seven-step process.

The Process of Answering Questions 207

Fig. 10.2. Question preparation sheet.

used if the question arises). Or finally, you could acknowledge the question but decline to answer because it is not relevant and refocus the audience on your topic.

Listen to the "Attitude of audience toward presenter during Q&A" podcast on the DVD

When you rehearse your presentation in front of a mock audience prior to the big event, plan to include a Q&A session as well. Make sure your mock audience includes experts and non-experts because their questions differ. Another way of preparing your Q&A is to review the standard questions one usually encounters during a scientific Q&A. They are listed under the last heading of this chapter.

Asking the audience. When the time comes for the audience to ask questions, you typically tell the audience something like *This ends my presentation. Spending this time with you to share our research was a delight.* This statement gets the attention of the chairperson who in turn thanks you for your presentation and asks the audience if there are questions. However, I prefer the next scenario in which you keep the chairperson out of the process and retain your role as host. Conclude, thank the audience, and immediately move toward the audience away from the lectern, smiling, arms stretched forward with palms up in an inviting gesture, and express your happiness to answer questions with a sentence like *I'd be delighted to answer the questions you may have on our work and on these conclusions.* (Remember that your conclusion slide should still be on the screen at the start of the Q&A.)

As soon as you stand there facing the audience, silently smiling, waiting for the first question to come, understand this: you are the host, you are not the victim facing a firing squad. Relax, look positively friendly and inviting, and WAIT. The pressure to ask a question has now been transferred to the audience. The people facing you are under pressure, not you; therefore, relax. Breathe deeply as you count slowly and gradually turn your body to face the whole audience from left to right and back, eyebrow raised, establishing eye contact briefly with people in the room, one by one.

Breathe in — one hundred thousand and twenty seven
Breathe out — two hundred thousand and twenty seven
Breathe in — three hundred thousand and twenty seven
. . .
Breathe out — nine hundred thousand and twenty seven

Give enough time for the audience to muster the courage to ask questions, and stop panicking that you might lose face if no one has questions for you. In the meantime, you are pumping a little more oxygen into your blood for the time your brain requires it.

Before the end of your counting, one or several hands will go up. If several people raise their hands, your role (unless the chairperson does it) is to clearly indicate from whom you will take the first question. Be specific: use your whole arm palm up to point in the direction of the person — no finger pointing please. And be courteous: avoid saying, *Yes, you there, at the back, with a green jacket*". Instead, say, *Yes, I will take the first question from the smiling lady* (at this time all gentlemen, and unsmiling ladies will lower their hands) *at the back* (other smiling ladies raising their hands at the front of the room will lower their hands and all will remember to smile when asking their questions).

If not one hand goes up after the ten unhurried seconds have passed, the chairperson may ask a question. That question may even be a written question you prepared for the chair prior to your presentation — just for his or her convenience, of course — you made that point clear. Alternatively, you may yourself ask a question on behalf of the temporarily frozen audience to give people a little more defrosting time. Such questions may be about your future work because there rarely is enough time to mention it during the talk. *Some may be wondering in which direction we have taken this research after discovering . . .* Once you have briefly answered your own question or the chairperson's question, renew your request for questions. If no questions are forthcoming, cut your losses. Do not thank the audience for their questions and do not make a silly unprofessional remark like *well, I must have been pretty clear since you all don't have one single question*. It simply reveals your embarrassment. Just tell the audience that you will be available outside the room if people have questions or want your printed colour handouts. You are ready to network.

After responding, if time (or the chairperson) allows, ask the audience again if there are more questions. If several people raise their hands, make a mental note of it so that at the end of your answer, you can return to the person whose question is still pending. Point (palm up) to that person, and say *Yes, sir*, or *Yes, Madam, you had a question, I believe*.

Listening to the question is the most difficult part of a Q&A. At this stage, you are usually tense and stressed for fear of not being able to understand or answer the question. Trust me: stress makes you deaf. The less you stress, the easier it is for you to listen effectively. Effective listening is totally dedicated to the understanding of the question so that you can rephrase it without difficulty for the benefit of the audience (and yours too). The anagram of "listen" is "silent". You have to silence that part of your brain that actively tries to guess the question. As soon as it thinks it knows, usually before the question ends, it starts looking for an answer. In effect, the brain steals listening cycles to allocate them to the search for an answer. Theft rarely goes unpunished! In a relay race, the runner who starts running too early before the baton can be safely passed is jeopardising his chances to win. In the context of a Q&A, the penalty is also serious. Often, the question is misunderstood, and the answer misses the mark. The audience, not as tense as you are, immediately notices it — none of this is good for you. The signs of ineffective listening are the following: 1) the presenter starts answering before the person asking the question finishes the question; 2) the presenter never takes the time to rephrase the question; 3) at the end of the answer, the presenter asks *did I answer your question?* (If the question is clearly understood, there is no need to check whether your answered it correctly!); and 4) the presenter asks the person to repeat the question when the whole audience has clearly understood the question.

To have a good answer, you need to fully understand the question. Therefore, focus on the question. Do not attempt to even think about answering it before you fully understand it. It is possible, however, that despite careful listening you do not understand the question. It may not be your fault at all, and the audience may be just as puzzled or unclear as you are. The person asking the question may indeed be unclear, or have a strong accent, or may have insufficient knowledge in the field to ask a clear question. To face such situations, you are not without helpers. You could turn to the chairperson and, using gestures, communicate your inability to understand the question. The chair would then take over the process and possibly ask for the question to be written down. Besides the chair, the audience is often quite willing to assist you and clarify or rephrase the question spontaneously when you clearly do not understand. Usually, a compatriote of the questioner whose accent is troublesome, someone with a better command of English, will rephrase the question in clear English. It is rare, however, to understand absolutely nothing. Often you will understand enough pieces of information for you to guess what the question is. In that case, simply rephrase what you believe the question to be, and indicate your doubt with words such as *If I understand you correctly* ...

Rephrasing the question serves many purposes.

1) It allows your questioner to agree or disagree with your understanding of the question — a misunderstood question leads to an unsatisfactory answer.

2) It gives your questioner a chance to clarify an unclear or ambiguous question by restating it using different words.

3) It gives you the assurance that the question is well understood when the questioner agrees with your rephrasing.

4) It gives you time to think of an answer while rephrasing the question.

5) It gives you the opportunity to rephrase the question more clearly and succinctly for the benefit of all.

6) It allows you to tag an optional comment before or after rephrasing the question to highlight its relevance and raise people's level of attention.

7) It allows you to slow down the rapid fire of questions coming from the same person.

8) It gives people in the audience who may not have heard the question properly, a second chance to hear it clearly through your microphone. Listening to an answer without knowing the question is very frustrating.

9) It allows you to rephrase a highly technical question using less technical terms for the benefit of the non-experts.

As you can see, rephrasing is essential — for you, for the audience, and for the person who asked the question. Be a gracious host. A presumed understanding of the question asked is not sufficient. Think of the people in the audience without a clear understanding of the question, and think of the questioner who may need your help to become clear. Rephrase.

Not rephrasing is only justified in the following exceptional cases: a) when the question is loud, short, spoken in perfect and well-articulated English, and you do not need time to think before answering it; b) when your foreign accent is ultra strong; c) when you are unable to even guess what the question is; or d) when you do not want the audience to hear the question again because you will decline to answer it.

Rephrasing is a technique you must learn. It is improper to repeat the exact words of the original question as in the following example. Imagine the question is *What is the temperature of melting polar ice?* Rephrasing is not repeating as in so *you would like to know what is the temperature of melting polar ice?* Instead, you would say, *At what temperature does polar ice melt?*

Even though you are rephrasing to a person, it is not always necessary to include that person in your answer with the "you" pronoun. In the previous example, the question was rephrased without a reference to the questioner. However, if you include the questioner, be personal and use *you*. Do not refer to the questioner in an impolite and impersonal manner as in *this man would like to know*.

Somehow, when it comes to rephrasing, presenters sound mechanical. They use the same sentence starter as in *so the question is* ...or *you would like to know* ...It is much better to vary the starting statements so that no rephrasing pattern becomes apparent over time. Here are a few starting patterns:

1) Start with a question word to rephrase the question: *who* ... *what* ... *where* ... *which* ... *when* ... *why* ... or *how* ...

2) Start with a qualitative assessment of the sentence (interesting, good, difficult). Follow with the rephrased question. Then say why you found the question interesting, good, and difficult.

3) Start with a thoughtful silence while you try to simplify or condense a long question into a shorter one, or a complex question into a series of simpler questions. *If I understand correctly, you are interested to know why* ... or *this question has two aspects: the first is why* ... *and the second is how* ...

4) Start with a brief comment reflecting the motivation behind the question (doubt, contradiction, etc) and then rephrase the question.

You have rephrased, now **validate the question**. Look toward the questioner for validation. As you rephrase always reflect your uncertainty if you are unclear about the meaning of the question. It is a direct indication to the questioner that you request a formal validation as in *If I understand you correctly* ...

Rephrasing and processing the question take place simultaneously. As you think of the words to rephrase the question, you also think how to answer it. Processing of the question continues after the rephrasing stage. There may be a brief pause during which you collect your thoughts or give a comment because extra time is needed

1) to ascertain whether the question is relevant or not (relevance is not always clear cut)

2) to figure out how to answer it succinctly and yet satisfy the questioner

3) to carefully choose words when answering difficult questions such as hostile questions

4) to think of safe ways to answer a reasonable question (from the point of view of the audience) for which you do not have a ready answer and

5) to understand the motivation behind the question.

Commenting. Apart from giving you extra time to think of an answer, the optional comment helps you transition smoothly into your answer. However, avoid flattering comments such as *A very good question indeed. Why ... Don't let that kind of flattering comment become a pattern.* If all questions are good questions, then no question is a good question. The audience will consider that statement to be pure hypocrisy and symptomatic of a lack of honesty on your part. *What is the boiling temperature of water* is not a good question. It is a trivial question. If the question asked is truly a good question, then say so because it will immediately raise the attention of the audience by raising the expectation of an interesting answer.

A comment, such as a possible reason for the question, can be part of the rephrased question. Imagine the question is *Since you add salt to the solution, shouldn't its boiling temperature rise instead of decrease?* The rephrasing could be *This is a shrewd observation. Why do our results seem to defy the laws of physics? After all, when you add salt to water, its boiling temperature increases. Am I right?*

However, do **not** reflect the apologies that sometimes precede a question. Some questioners will say, *I am sorry, I do not understand why ... Since you* are a respectful host, you will NOT say, *You are unclear ... You do not understand ... * or *You fail to see why ...*

Use the anonymous passive voice *it is unclear why ...* or *why ... is unclear.* Even though a lack of clear understanding is often due to your own lack of clarity during the talk, or it may not be your fault at all. It may just be that the person who asked the question missed one part of your talk (late arrival), or was distracted while you were explaining. It is tough to sustain attention over a long period. Be gracious to your audience but do not take the blame for the questioner's lack of clarity. There is no need to humbly apologise with a sentence like *I'm sorry. I have not been clear.* Mentally take note that you may not have been clear and the next time you give that presentation again, make sure you thoroughly rehearse that part of your talk, or clarify the slide that caused the question.

Well used, a helpful comment reveals more of your personality in many ways:

1) By demonstrating that you have great analytical skills, as in *there are really two parts to your question ...*

2) By recognising the value of a question, and therefore, the value of the questioner, and your own value as in *This question had given sleepless nights to many of us in our team before it could be answered.* Such comments usually raise the attention level in the audience. Be aware of another potential negative impact of the catchphrase *This is an interesting question.* The sensitive people who have asked questions prior to this interesting question and have not been fortunate enough to be recognised for their *interesting* questions may think that you did not find their question interesting. That is why it is always appropriate to justify why the question is indeed interesting and stands out from the rest.

3) By explaining the question in logical terms, projecting your logical skills as in *This is a valid question. If the temperature is not a contributing factor to the behaviour of this chemical, then why is pressure, since the two are related?* (The question may have been *why isn't temperature contributing to this behaviour?*)

4) By showing that you value the opinions or suggestions of the audience as in *Thank you so much for your helpful comment.* Take a card from your shirt or trouser pocket (you placed it there before the talk) and make a quick

note of the suggestion in front of the audience, thus showing that you indeed value their suggestions. Actions speak louder than words.

Rule 1: Answering should always be brief. **Rule 2**: Any question that has been heard by the audience must be answered. These are the only two rules to strictly follow. Your intent is to answer the largest number of questions from as many people as possible in the time imparted for the Q&A. If the answer requires time to give full satisfaction, just state the main points. If necessary, make sure to tell the questioner that you will be available after the talk to provide a more comprehensive answer.

Why should the speaker answer as many questions as possible? It is because each question is revealing and helpful:

1) It reveals who is interested in what you do (precious information for preparing networking).

2) It reveals what is of interest in your talk.

3) It also reveals potential problems in your presentation such as lack of clarity, lack of evidence to back up your claims, lack of additional technical or background slides to answer questions, or lack of question preparation.

Even the lack of questions is revealing.

1) The presentation was too shallow. The audience has not learned anything it did not already know.

2) The presentation was too dense. The audience is overwhelmed and does not know what to ask because it understood little.

3) The presentation was too self-centred on the needs on the presenter, with little regard toward the needs of the audience.

4) The audience really did not understand the presenter's strong accent and does not wish to struggle through oral questions.

5) The audience does not like the presenter: the way he turned his back on people and constantly faced the projection screen, his reading of all text on

all text-heavy slides, the way he avoided eye contact and crossed arms or walked like a peacock or put both hands in his pockets, or kept his cap on …

6) However, it may be also because the chairperson does not allow the presenter time for questions either because of an overlong presentation, or for reasons outside of his control.

7) This list may be demoralising but get comfort with the thought that the presenter is not always responsible for the no-question situation. For example, he may be the last presenter of the day and the people want to go home after a gruelling day of presentations. He may be in front of a very shy audience, one whose culture values those who do not ask face-losing questions in public and who prefer private questions in one-on-one conversations after the talk.

Apart from being brief and precise, answering should always be courteous. You are the host. Even if a question seems trivial or if you are convinced that its answer is obvious (had the questioner paid attention to what you already said during your talk), you cannot show your irritation. Remain the smiling host and pretend that the question has not been answered during your talk even if both you and the rest of the audience know it has. The way an uncourteous host reveals irritation is quite subtle. One may look away from the questioner and answer in a toneless voice, as if bored from repeating the same thing. Another one may say the horrible words *As I said before*, or *As I have shown before* (and proceed to display the slide with the requested information). This is a slap in the face to the person asking. It implies that they did not listen or pay attention.

Aside from being brief, precise, and courteous, answering should also be honest and explicit — avoid *you know what I mean*. If you do not know, you do not know. Do not make up a story, or venture into the minefield of the possibly right or wrong answer path. Scientists easily detect half-truths. They are good at pushing an argument to find out whether you know or not, and whether you are sure or unsure through one or two insightful serial questions that will immediately expose your false claims and lack of knowledge. You would then lose all authority as a scientist and be exposed to ridicule in front of your peers. If you do not know, it is not necessarily bad. The question may not be relevant; it may be outside the scope of your work. Declining to answer a question on the grounds of lack of data or lack of expertise is quite acceptable. It may be an open research issue.

Finally, answering should always be an opportunity to strengthen the value of your contribution. Some of the answers may be negative: *We haven't done that*. Whenever you end an answer on a negative note, and especially if many such answers end that way (*we haven't done that either*), you destroy the positive perception of your contribution created during your presentation. It is vital that you end your "negative" answers on a positive note, reinforcing what you have achieved or highlighting the applicability of your work to other situations.

Networking is so essential when one does research. Future papers may be written jointly with some of the people sitting in front of you in the audience. Financial grant support may be within reach through the intermediary of a decision maker sitting five rows away from your lectern. Your future collaborator or your next boss may be right there attending your talk. It may even be that someone's insightful comment will redirect the whole course of your research for the next five years.

The audience responded to your invitation; they are your guests. You want to know your guests better, and therefore you have prepared something they can remember you by: handouts maybe, a colour copy of your presentation on a CD-Rom, your business card, the address of your website, your email contact… This is going to be your parting gift. Naturally, you will keep yourself available outside the meeting room (or besides the podium if the next speaker is scheduled in 10 minutes) to answer further questions and distribute your handouts in exchange for their business card or email address. It takes two to network. The benefits of networking are not always equal; one does not always benefit as much as the other, but networking is not a matter of winning or losing, it is a matter of being connected to what happens and to who makes it happen. The benefits of networking are not always immediate. Some networking contacts may be activated months after they were set.

Three Troublesome Questioning Styles and How to Deal with Them

Here is a dream scenario. You just asked if there were questions. You get three single questions from three different people. Each question is friendly, clear, short, and can be answered succinctly to the entire satisfaction of the questioner. Don't you wish all Q&As were like that? Alas, along come the rambler, the serial Q&A killer, and the enumerator to make your life difficult.

The question or comment from the rambler never seems to end. Its topic seems to change as time progresses, leaving you uncertain as to what the question really is until the very end … or never. The problem with rambling is that it shortens your Q&A time, and therefore your chances to identify people valuable to you, or people you could be valuable to.

The serial Q&A killer is not immediately recognised. The question is normal, but the single question turns out to be the first question. As soon as one question is answered (sometimes while it is being answered), another question follows. The questioner absorbs all available question time, leaving no opportunity for others to ask questions. This may look like a blessing from your point of view — after all, you are getting many questions — but in fact, it is a curse. The audience feels left out of what becomes an elongated dialogue between two people only — and you forego the opportunity to identify other interested participants.

Contrary to the serial Q&A killer, the enumerator announces all questions at once. The enumerator wrote many questions while you were presenting. Well-organised and methodical, the enumerator has written down the questions for fear of forgetting them, or to make sure they are written in good English and can be spoken without hesitation. Some of the questions may have a conditional statement placed at the head of the question such as "if this … then [first question] but if that … then [substitute question]." The problem with such questions is that you do not have a piece of paper on which to write the questions and your memory is not good enough for you to remember all of them without having to ask the enumerator to repeat them. Furthermore, answering all the questions takes time, and other people do not have the opportunity to ask their questions.

How does one deal with ramblers?

When ramblers start talking, you do not know that they are going to be ramblers, nor does the audience. Some questioners appear to be ramblers, but they are not. They are simply people who need time to express their questions. Maybe they think of the question as they speak it and require several rephrasings before they get the question right. Maybe they struggle with words because they are not expert in your field. The question is long, but you do feel that a question is forthcoming.

Ramblers are not there to ask a question, but to be noticed by the audience, as experts greater than you, or as people of importance. In other words, ramblers are on an ego trip. Since they want people's attention, they may start with a comment, usually a long one and occasionally a friendly one to lower your guard, and finally come up with a question — but only if you force them to ask one. Occasionally, ramblers hijack your presentation and use your Q&A time to present their unpublished (or published) work. Ramblers do their networking at your expense during your session, and you have to stop them.

How do you do it while not appearing rude?

1) As soon as you feel that a questioner is a rambler (and the audience probably feels it in the first ten to fifteen seconds of the rambler's speech), find a way to agree with a statement from the rambler and then turn that statement into a question. *Your question is probably whether or not this is always true, or are there any exceptions, is this correct?* As soon as you posed that artificial question and asked the rambler whether this is the question or not, the rambler has no other choice than to formulate a question.

2) You may turn toward the chairperson of your session while the rambler speaks. This silent body gesture may be enough to prompt the chairperson into action.

3) Since everyone has to breathe, you may want to look for a slight pause in the rambler's speech, and then loudly and with a beaming smile, thank the rambler for his comment and immediately indicate that you are happy to take the rambler's question. The smile makes a big difference: you are the friendly host, and you are in charge.

4) Hold out your arm horizontally, palm open facing the rambler (a stop sign), smiling, until the rambler stops talking, express your thanks for the comment and ask, *Do you also have a question?*

How does one deal with serial Q&A killers?

The term "killer" seems a little strong, but this type of questioning effectively hijacks the presenter as the Q&A process turns into a dialogue between two people; a dialogue that removes all possibilities for other people to ask their questions. In a short two- or three-minute Q&A, the serial Q&A killers will consume all the time there is left. They have to be stopped. There are two types of serial Q&A killers: friends and debaters. Friends are genuinely interested in your work. It may be because it has immediate impact on their work, or because you greatly arose their curiosity. Since your presentation is inevitably short in details, there is much left to know, and the friendly questioners want it all, there and then. Debaters, on the other hand, are out to win an argument, whatever it takes. The flow of questions continues until they win. Sometimes debaters argue for the sake of arguing, just to be perceived by others as researchers with great critical skills. Sometimes, they bring the audience into their debate and leave the presenter out of the debate, alone, helpless behind the lectern, wondering how to regain the attention of the audience. There are a number of ways to stop serial Q&A killers, but not answering their question is never the right way.

1) Remember the role of the chairperson. When the serial Q&A killers come with their next question, before answering, you could turn toward the chairperson and ask if there is time to answer the question, thereby signalling that time is not under your control. A chairperson familiar with serial Q&A killers may ask you to answer briefly and return to the audience for other questions, a clear stop sign for the serial Q&A killers.

2) In the previous case, as in this case, the tactic is to pre-empt the next question by warning the serial Q&A killers that after answering this "last" question, you will turn to the audience for their questions. How this pre-emptive strike is worded differs according to the type of Q&A killers. Thank friends for their great interest in your research (smile), and say that you will not be able to exhaust the many interesting questions they have during the session. Invite them to resume questioning at the end of the session. Tell debaters that you appreciate their concerns over the way your research is conducted and invite them to continue the discussion after your Q&A.

   For both types of serial Q&A killers, keep your last answer short. At the end, DO NOT turn toward them (or ask if you answered their question!) because they would see it as an encouragement to continue despite your warning. Turn AWAY at least 45 degrees from the serial Q&A killer and tell the rest of the audience *I am now also happy to answer your questions*.

3) It is easy to fall into a dialogue — a one-on-one conversation with serial Q&A killers. To find ways to treat the problem, let's observe the symptoms. First, the pace of questioning is fast and the questions short. Second, serial

Q&A killers maintain constant eye contact with you, and expect you to do likewise. You therefore need to slow down the rapid fire of questions and reduce eye contact. First, simply take your time to rephrase, even simple short questions, giving you a breather and an opportunity to pre-empt other questions. Second, turn away from them when you answer, face the audience, and do not return to them at the end of your answer. It gives you a chance to see if other people with questions want your attention (they usually raise their hands even as you continue talking). Let your body language do the work of indicating that you no longer want questions from the serial Q&A killer.

Finally, how does one deal with enumerators?

Out of the three types of troublesome questioners, enumerators are the easiest to deal with ... as long as you are prepared. To face enumerators, one needs a little accessory kit: a rigid blank card and a pen or pencil in the shirt pocket or at hand. The card has to be rigid because you will be writing on it while standing up facing the enumerators. The card is there for you to write down the questions so that you do not have to ask the questioner to repeat them (unless your memory is phenomenal). Here are three possible tactics for dealing with enumerators.

1) Look at the chairperson when the list of questions has been read. The chairperson may indicate that you have time to answer only one or, at the most, two questions.

2) Listen to the questions while taking notes. If some of the questions require more time than you have in the Q&A to answer, just say so and skip them. If some of the questions are troublesome — either they are not relevant to your work, or you don't have a good answer for them — do not say so, but do not answer them. Instead, select the questions that have a quick and easy answer, or that are particularly interesting, and then tell the enumerator, that in the interest of time, you will only answer question x and y. You are in charge; you are in control. This is a great way to avoid problems. The card may even come in useful in other situations such as a question to which you have no immediate answer but you offer to respond through email. If you say *I'll get back to you on that*, people will know you do not intend to do so. But if you actually write down the question, and say that you can later email an answer, people will appreciate your willingness to

help and will make sure to give their coordinates after your talk. It may even be that someone is giving a suggestion, and making note of it live on the spot indicates your interest and willingness to consider it instead of giving the standard *this looks interesting. Thank you very much. I'll be sure to look into it*, which will probably be interpreted as "he is just being polite but not really interested".

3) If the enumerator announces three questions but stops at the end of the first question expecting your answer before asking the second question, you should ask for the rest of the questions. Alternatively, if you know the answer to that first question, before answering, ask the questioner for the permission to answer the other questions right after the Q&A.

Difficult and Dangerous Questions

Most of the questions a presenter gets from an audience have one particularity: they are directly related to what has been shown on the slides, or to what has been said. This is indeed a sign that a significant part of the audience is not familiar with your topic. Non-experts can only ask questions on what they see and hear because they don't know anything else! The experts, on the other hand, while also asking questions on what is presented, will add their own pointed questions to get more detail. Keep that in mind when you prepare your questions from your simplified slides — those which underwent a *data-suction* treatment to remove details. Make sure that data-rich slides follow your conclusion slide to satisfy the expert's appetite for details.

Questions from Experts

Expert questions are needlessly feared by you, the presenter. Nobody knows as much as you do on your topic. As expert, you are confident you can answer all relevant questions. You may however have problems with the questions that fall outside of your comfort zone.

Since you are not familiar with the questioners' work, you do not know what in your work overlaps with their work. Similarly, since the questioners are not familiar with your work (a presentation is always too short), they can only guess where your work overlap theirs. In Fig. 10.3, the experts believe your work to be within the dotted line circle when in reality it is within the

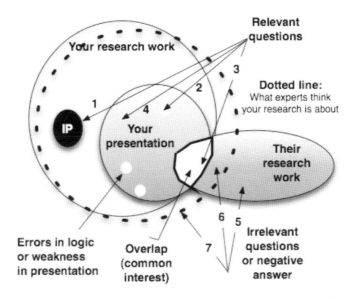

Fig. 10.3. Taxonomy of expert questions.

full circle. Consequently, there will be a misperception of the overlap area. The intersection between the oval and the full circle is not the same as the intersection between the oval and the dotted line circle.

Experts are eager to see commonalities where there aren't any. Call it "wishful thinking" if you want, but you may be the cause of it. You may have oversold your real achievements, or they filled up your lack of details with their incorrect assumptions.

On Fig. 10.3, seven types of questions are identified. They belong to two groups. Questions 1 to 4 are relevant questions, and questions 5 to 7 are irrelevant questions.

- Question 1 targets the undisclosed IP (Intellectual Property).
- Question 2 investigates the possible holes or undisclosed information in the work presented (logical issues, undisclosed limits or assumptions, errors).
- Question 3 probes the area of work that is common to presenter and audience.
- Question 4 is a normal question (details pertaining to what you presented).
- Question 5 brings you onto the research turf of the questioner.
- Question 6 falls outside your area of expertise even though the questioner thinks it falls within your area of expertise.
- Question 7 attempts to discover the boundary and scope of your work and falls outside of it.

These question types fall into several categories: The come-onto-my turf questions (5), the boundary questions (3 or 7), the what-if or have-you-tried (3, 5, or 6) questions, the poke-a-hole (2) questions, and the IP questions (1).

When the time for questions that comes, you can be asked relevant questions with affirmative or negative answers, or irrelevant questions. Irrelevant questions are not meant to be irrelevant; they just express the questioner's attempt to see whether your work is applicable to their research, or whether their work is applicable to your research. Since you do not know their work, or at least not with the knowledge depth necessary to answer these questions, you should not attempt to answer them. These come-onto-my-turf questions are irrelevant, and unanswerable (unless you happen to be an expert in the questioner's field too).

Therefore, when you listen to the question, ask yourself this: am I competent to answer this question? If you are not, and are not expected to be, do not answer it. For example, imagine the question is *how does your hydrogen fuel cell system compare with a typical solar cell system in terms of efficiency?* You are not a solar cell expert. You have read about solar cells, but this is not the focus of your research. Some presenters think they lose face if they are not able to answer a question, and propose an uneducated answer. If the answer is incorrect, they lose credibility. Therefore, do not step outside of your comfort zone. You know you are venturing on dangerous grounds as soon as your answer starts with *I think that … I believe that … or I suppose that …*

To answer the **come-onto-my-turf** questions, you could do the following:

1) Recognise that both your field and the other field have a common purpose (here, the search for alternative sources of energy), but decline to answer because you are not an expert in the field mentioned in the question (here, solar cells). End your answer by restating the efficiency reached by your method (the hydrogen fuel cell), so as not to finish on a negative note.

2) Suggest to the questioner that criteria for comparing these two technologies (solar and hydrogen-based) should first be established, and follow through by restating the factors that you have used to determine the efficiency of your technology. In effect, you are claiming inability to answer without first doing a preliminary task. But you are also demonstrating your willingness to entertain the question.

3) Say you do not know the answer to that question, but propose a closely related alternative question that you can answer. For example, say that you can compare the hydrogen fuel cell efficiency today with the expected hydrogen fuel cell efficiency tomorrow, and mention the efficiency-limiting factors whose importance is likely to decrease. In effect, you are repositioning the question on what you know.

Experts also use **boundary** questions because they want to ascertain whether what you have done would also apply to a slightly different situation that they are facing. In other words, they are poking at the boundaries of your work to see whether they can be extended or not. For every diagram, there is a lower limit and an upper limit to the range of values. You decided on them. Can you justify why you did not go beyond them? Is it due to the limitations of your measuring instrument? Your experimental set-up? The purity of the chemical you use? The availability and cost of some component in your experiment? ... These boundary questions are very relevant and the audience will expect you to know the answer if you are an expert. It is therefore essential to look at every graph in your presentation and ask yourself these scope or boundary questions. There are no techniques here to learn. Just make sure you know the answer to these kinds of questions prior to your talk. If you do not know the exact number, give an order of magnitude.

Another type of question asked by experts is the **what-if** question. It is equivalent to the **"Why-didn't-you"** or **"have-you-tried"** question. The question *What if you had used the so and so method?* is the same as *Why didn't you use (or have you tried) the so and so method?* The aim of these questions is to assess whether or not you have explored (and are knowledgeable about) alternatives. These questions can be relevant or irrelevant, so be attentive. *What if a new silicon alloy had been used instead of Gallium arsenide? What if Hidden Markov Models had been used instead of Neural Nets?* Surely, your materials and methods are perfectly suited for the task, right? If you have chosen a particular method, it is for a specific purpose, and therefore you should be able to present the reasons for your choice. If the suggested alternative method/material/formula ... is well known in your field, the audience expect you to justify your choice. It is a fair question. Questions asking you to justify the simplification of a problem by reducing its dimensions are also fair. For example, you may be asked, *Why did you propose an algorithm that handles 2D data when in reality the problem you are tackling is clearly a 3D problem?* You'd better have a more satisfactory answer than *it was easier this way, or it would require too much computing power!*

Some **what-if** questions are irrelevant: *What would happen if you applied this method on this metal?* They don't have to be answered. However, you may make a mental note of the question and network with the person who asked it, after your talk. There may be something about that metal worthy of interest.

How do you answer **what-if** questions?

1) Be prepared to justify your choices on scientific grounds. The audience may find some answers hard to swallow, such as *it was simpler not to do it*, or *it was too hard to do it that way*, or *our team does not know how to do it any other way*, or *everybody does it this way*. The audience feels that you have taken shortcuts instead of tackling the hard problem, or that you use a Neural Net to solve a problem because you are a Neural Net expert. Here are examples of more acceptable answers:

   a) *We kept the model simple from a computational complexity perspective, and yet still achieved good agreement with the experimental data.*
   b) *We considered that variable at one point but chose to disregard it because its P value was too high given our dataset.*
   c) *We used HMM instead of NN because we noticed that, given a data point, the next data point seemed more probabilistic than deterministic.*

2) If you have much time for Q&A, you can return the question to the sender in question form: *um, interesting, may I ask why would one want to use method XYZ?* Either the embarrassed questioner responds: *I don't know, it was just an idea*, and you are still seen as an expert, or the helpful questioner makes an unexpected and interesting comment. In the last case, you would thank the questioner for the interesting suggestion but highlight the particular advantages of your own method; alternatively, you could suggest collaborating on a joint paper around this interesting angle. In both situations, you remain an expert.

3) State that you have not considered that alternative because your method seemed very promising, and give the reasons why it was promising.

Among the expert questions, the most difficult are the "**poke-a-hole**" questions (debating serial Q&A killers often use poke-a-hole questions). With these, the expert intends to ascertain where the Achilles' heel is in your research. It could be a limited scope, a limited dataset, a bias in the data set, a faulty process or procedure, an unreasonable assumption or a crippling one that severely limits the usefulness of the findings. Sometimes, the questioner, on an ego trip, wants to demonstrate his seniority over the presenter, and sometimes the presenter's misbehaviour triggers such poke-a-hole questions. Some presenters show lack of respect toward other people's research by displaying comparative tables that put down everybody's work except their own. Some over-claim without due evidence. Some are blatantly arrogant. Some betray their shallow knowledge through glaring erroneous statements. Some make glaring labelling, grammatical, or spelling mistakes on the slides. Finally, some fail to mention the source of "borrowed" data or visuals, thus plagiarising.

How does one deal with **poke-a-hole** questions?

1) Prevention is better than cure. Prevent these questions. Have experts (your manager and peers) check your presentation prior to your talk.

2) Be prepared to justify your data, your method, your results, and your conclusions. The only way to deal with poke-a-hole questions is to have your arguments shielded and armoured with solid data. Keep extra slides after your conclusion slide to help you answer the poke-a-hole questions.

3) Give a brief answer indicating that you understand the question. You may flash a technical slide, but do not dwell on it: the expert reads it faster than the non-expert, and that slide is for the expert anyway. Offer to give further details after the talk. Well answered, poke-a-hole questions establish and showcase your expertise.

The IP question comes from an expert. To protect intellectual property, one can adopt one of three strategies: patent it, keep it a trade secret, or publish it. Publication is equivalent to putting the IP in the public domain. It does not provide protection to the same degree a patent or a trade secret does, but it establishes prior art (a trade secret does not) and prevents anyone from blocking you with a patent on what has become prior art. When your research institute, company, or university allowed you to present your material, it made sure that their secrets and patentable information were not part of your presentation. Along comes a wolf in sheep's clothing to your presentation. S/He is Mr. or Mrs. Nice, friendly, full of praise for your work, congratulating you, nearly making you blush with pride. Then come a quick set of two questions. The first question or comment usually brings you in the vicinity of the IP, but not right in it. The second question is short and innocent but right in the core of the IP. A final one may just plant the last nail in your coffin. For example, let us say that you have found a non-destructive process to assess whether or not the typical glass pane one sees on skyscrapers might shatter in certain conditions (potentially killing innocent pedestrians walking underneath). You wrote an invention disclosure, and the lawyers in your research centre are in the process of drafting a patent for a yet-to-be-determined number of countries. Your paper is on the thermal effect of ultrasounds on tempered glass panels. As it turns out, your lab is working with a glass manufacturer and you have identified (here is the patentable material) that for a given ultrasound frequency mix at a given intensity, a well-defined glass impurity causes glass to break. The ultrasounds, by causing faster thermal expansion of the impurity than the surrounding glass, make the glass break. Fortunately, the local temperature increase in the area of the impurity can be detected by a sensitive infrared photo of the whole pane of glass subjected to the ultrasound frequency mix. In other words, you can predict problems effectively and cheaply — but this is not part of your presentation. Your paper does not describe that process. It simply presents the thermal effect of ultrasounds on tempered glass panels.

Here comes Mr. Nice with his comment and questions. He already noticed that all your glass samples were 1 cm thick, and made of tempered glass and therefore, probably used in construction. *Thank you very much for this excellent presentation. I could not help but notice that, in one of your graphs, the standard deviation for the heat measurements is greater at some frequencies. What could cause the increased variability at these specific frequencies?* You carefully listen, and you rephrase the longish question: *Why does the temperature curve display little heat bumps around specific frequencies?* You then answer, *It could be from a number of reasons; for example, random variations in glass thickness, or random variations in the intensity of the ultrasound at random frequencies.* You pause a second before saying *or random variations in glass purity, I suppose.* You said that last part on a lower tone. The IP seeker noticed it. He is smart. *Surely, the effect of variation in glass thickness or in ultrasound intensity is equally spread across all frequencies, isn't that correct?"* You are now cornered and have to admit that the most likely cause is variation in glass purity. Naturally, such an answer reveals that you are a real expert, but it also decreases the number of the claims you could make in the patent. You just made public that, at defined ultrasound frequencies, impurities in glass raise the local temperature of the glass enough to be measurable using infrared scanners.

How does one not fall into the trap of the IP question?

1) Avoid displaying data that betrays your IP (unless it is already patented). In our example, the experiments could have been conducted on inclusion-free tempered glass. On the curves, the telltale sign would no longer appear, yet the conclusions would remain unchanged.

2) Be suspicious of flatterers, listen attentively to their questions and look for their motives. Historical reconstruction questions are good ways to identify hidden IP. Here is one. *Congratulations! This is a great piece of research. What are the particularly hard problems you met during your research and how did you solve them?*
   Answer that there were far too many to recall, but that they are now all behind you and that is where you want to keep them. Here is another IP seeking question.
   Excellent presentation! I am sure that you discovered many things of interest on the way to the results you presented today. Could you mention some of these secondary findings that may not have been presented here today?
   Remain evasive in your response, and mention a minor one far from your IP. Of course, if you don't have IP issues, do respond to these interesting questions but keep your answer brief to give the rest of the audience a chance to ask their questions.

3) Some IP seekers (or haters) are straight shooters: they directly ask whether anything in the research presented has been patented. Researchers do not like the type of contribution they cannot use without infringing on someone's patent, or without having to give away all derivative rights. A simple answer, one that preserves the goodwill of the audience toward you, could be this: *Whether there is or not, this question really should not be answered by me, but by my research institute. I'll be happy to put you in touch with them. Any other questions please?*

4) In the case of our tempered glass example, here is a possible strategy to answer the question *What could cause the variability at specific frequencies in these glass samples?* — First, you would not rephrase the question in order to detract attention from it, and you would not say it is a good question. Then you would claim, *We also noticed and have this on our list of topics for*

230 The Answerable Scientist

investigation. It does not directly affect our overall results, however and you would restate your main results.

Questions from Non-Experts

Non-experts mostly ask easy clarification questions, but occasionally they ask troublesome questions such as the senseless question, and the teach-me-the basics question. On Fig. 10.4, these irrelevant questions are identified. Note that the mismatch between the dotted line circle and your research work circle is much greater for non-experts than for experts. The non-expert only has a vague idea. This is why it is essential that you clearly define the scope of your work during your talk.

- Question 8 is asked by the non-expert to gain background technical knowledge.
- Question 9 is completely outside of bounds because the questioner does not really understand the presentation.

Why do presenters sometimes get these **"senseless"** questions?

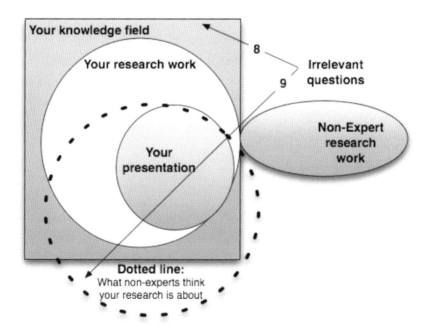

Fig. 10.4. Taxonomy of non-expert questions.

Such questions may arise from a misinterpretation or mishearing of what is being shown or said, or from a lack of knowledge on the part of a confused questioner obviously attending the wrong presentation. The questioner is so far from your field that the question is almost meaningless. To take an extreme example, imagine someone has been sitting through a presentation about Mendel's law of heredity. That person asks: *What other laws did Professor Mendel write?* This is bound to have the presenter pause for reflection a moment (and get a few chuckles from the audience). Strange things do happen to a stressed presenter who believes that all questions are relevant. The confusion is not immediately clear, and the presenter freezes as his brain struggles with the question. The presenter is even more frustrated because some in the audience seem to "get it". What is the appropriate thing to do?

1) Supposing you do understand the cause of the confusion, you are not to laugh or be irritated. You are the gracious host and do not laugh or show frustration at the expense of your guests. You simply explain in polite terms that Mendel was a botanist monk whose work contributed to the discovery of three principles or laws in genetics: the law of dominance, the law of independent assortment, and the law of segregation. There is no need to correct the questioner by saying that Mendel was not a lawmaker. And there is no need to thank the questioner for a good question. At the end of your short answer, ask for the next question. People will appreciate your tact toward the questioner.

2) If you are so caught with your topic that you do not see the confusion, asking to repeat the question may not take you out of trouble, and rephrasing it will only give you the approval of the questioner. You may therefore want to turn to the chairperson for help.

3) If the chairperson is obviously trying not to laugh, and motions you to continue, just say that you do not know how to answer that question, but you are willing to attempt answering the question again after the Q&A session.

How does one deal with the non-expert's **teach-me-the-basics** question?

The **teach-me-the-basics** question comes from someone who is very interested in your topic but is overwhelmed by lack of knowledge. Imagine your presentation is on the impact of various particles, from cigarette ash to hair dandruff, dead ants, and nail clippings, on the mean time between failures (MTBF) of computer keyboards. The question is *How do you calculate an MTBF?* At this point, some experts in the room may be clearly annoyed and look at the questioner with contempt. But you are the host. Do not launch into an extensive primer on the MTBF involving the expected value of the failure density function and the mean time to repair. Give a brief answer and move on to the next question.

Six Difficult Questions Regardless of Audience Expertise

There are troublesome questions that come from experts and non-experts alike: the **unintelligible** question, the **elaborate-on-conclusions** question, the **opinion** question, the **time-consuming** question, the **you've-got-me** question, and the **hostile** question.

Almost everyone knows a scientist with a particularly strong accent, soft voice, or speech impediment. At first, people struggle, and then, over time, they get used to the strange pronunciation and no longer consider it a problem. Unfortunately, in a presentation, while taking questions, you have no time to get used to the questioner's speech characteristics. If you are struggling, it is likely that others are struggling too; therefore, you need to rephrase the **unintelligible** question. Here are three techniques you could adopt (remember that you are the host!):

1) You have understood some words, missed a few others, but think you may be able to guess what the question is. While looking at the questioner for approval, rephrase your guess. If you are wrong, the questioner may spontaneously repeat the question using different words that you may understand better. Rephrasing is better than having to say, *I'm sorry. I do not understand your question. Could you repeat it please*, which shames the frustrated questioner because he knows that his speech problems act as a barrier to your understanding. As a host, this technique is gentler than asking the questioner to repeat the question without first attempting to rephrase your best guess. After the question is repeated a second time, if you still cannot understand the question, you could put on a puzzled look on your face, possibly clutch your chin, frown a little or shake your head slowly to indicate to the audience that you are still unclear. What usually

happens then is that someone in the audience will come to your rescue and translate the question into good English for you.

2) If you cannot even guess what the question is because the accent is too strong, you could simply turn toward the chairperson, using silent body language to indicate your inability to process the question. Once the chairperson is in charge, you can relax: the problem is out of your hands. Either the chairperson will ask the questioner to write down the question, or the chairperson can ask the person to repeat the question or ask the audience to help with the question.

3) If there is no chairperson, you could directly ask the audience for help with this question, as in *audience, I need help with this question. Could someone rephrase it for me?* If nobody in the audience is able to help, indicate your determination to answer the question and ask the questioner to kindly write it down and bring it to you. In the meantime, ask the audience if they have other questions.

With the **elaboration-on-conclusions** question, the audience assesses the level of confidence you have in your own recommendations or interpretations. You presented your results and gave your suggestions on what they possibly mean but you may not have told the audience why you believe so, and how you came to that conclusion. Alternatively, you may have told the audience, but some people could not follow your reasoning. The elaboration question is a relevant question, and a question you have to answer. It is similar to the following questions: *On the basis of result x, you concluded y. Is this sufficient to come to this conclusion? And are there alternative interpretations?* or *You claimed or stated that [...]. Would you care to elaborate?* If you answer this type of question satisfactorily, the audience will consider you an expert.

With the **crystal ball** question, someone in the audience wants to tap on your expertise to collect insights on future events. They really want what people call "an educated guess". The question may or may not relate to your work. If it relates to your work, it can be answered in the framework of your future works, you can remain authoritative, and you should answer using a formula popular in France. French people say, *La question arrive un peu tôt*, which means, *this question comes a little early*. Claiming this indicates that you do not have a ready answer but that, in time, you will be able to answer. For example, imagine that your work is still at the early stage and a question comes regarding its application in industry. You do not want to say that it has no impact, but you cannot yet give an idea of the impact; truly, this question does come a little early. Explain why. However, when you explain, do not place your work in a negative light: do not say, *We haven't yet done this, and we haven't yet done that*. Present your work as moving forward: work not yet done (negative) is work already in planning (positive).

If the crystal ball question does not relate to your work, your opinion is just as good as the opinion of the man next door. In that case, start your answer with a legal disclaimer stating you are not an expert and this is only your opinion; however, provide the reasons on which you base your opinion — remain a scientist. Alternatively, you could refrain from giving an answer altogether (this is safer), and claim that much depends on how things evolve in this or that area. Crystal ball answers always contain justifications.

The **time-consuming** question is one which, to answer properly (to your satisfaction and that of the audience), requires more time than you have during a typical Q&A session. Say so to the questioner. Provide the main elements of your answer, and refer the questioner to someone's work, or propose to postpone the rest of your answer until after the talk.

You are the host and want to make sure your audience is satisfied with your answer. If you know that your answer is unsatisfactory, say so, but explain why without apologising.

The **you've-got-me** question is a fair question that the audience expects you to answer because it is totally relevant to what you presented, but you do not know the answer. You stand to lose much credibility by revealing that you do not know. Why wouldn't you know?

Case #1. You are replacing the expert. The first author of the paper who was due to present the paper at the conference could not do it and you were asked to substitute for him. Your knowledge is insufficient to answer certain technical questions. Unfortunately for you, such a question comes. What do you do?

Case #2. You copied and pasted information from other sources into your slides. Your slide shows a photo without a source, or without a scale (when there should be one). You are asked about the source or the scale. You do not know without going back to the original (which you don't have). The photo refers to an acronym. You are asked for its meaning. You do not remember.

What are people doing in these situations?

1) Some find a way out by claiming that it is an interesting question requiring much time to answer correctly, and offer to answer after the presentation. This is not a truthful statement. Even if it saves you once, it will not save you twice. Furthermore, it does not work with case #2 situations.
2) Some lie and make up an answer while looking very confident, hoping people won't notice their answer is groundless. However, the audience does notice and the liars lose all credibility (for a long time).

As you see, there is no easy way out of such questions, but there are ways to limit their damage. There are two ways of coping with the you've-got-me question. You could admit that you do not know, and ask for the coordinates (phone or email) of the questioner to answer him as soon as possible, even in the next 24 hours (and actually do it). Alternatively, you could pre-empt such questions by announcing right before your talk that you are substituting for the first author and that you may not be able to answer all questions. State that you will take note of all questions you are unable to answer and have the first author answer them by email. At least, people will value your honesty and the effort you make to satisfy them.

Finally, here is the question everyone dreads: the **hostile** question.

The questioner may actually not be hostile at all, but in the manner the question or comment is put, you perceive it as hostile. Someone either disagrees with your findings, or turns a scientific issue into a moral issue such as a lack of respect or sensitivity. Whatever the situation, you feel attacked. How do you respond?

The techniques for handling seemingly hostile questions are based on your role as a host. A host never lacks respect toward guests. A host remains friendly at all times. A host does not take criticism personally but considers criticism as an opportunity to explain and clarify. A host never blames the guests. A host defends accomplishments so as not to lose authority, but does so on solid scientific grounds — not by lying, or by claiming that everybody does it this way, or by expressing doubt about the validity of the questioner's claim.

The process of answering hostile questions includes additional steps.

1) It is always necessary to rephrase a hostile question, even if it is clear. The rephrasing includes one extra element: the feeling, opinion, or sentiment

of the questioner. Having such feelings recognised helps the questioner calm down. People who are upset want both their hurt and their question to be recognised. Simply dealing with the question while brushing aside the anger is not sufficient. However, in your rephrasing, avoid mentioning yourself as in *It seems that the way I presented this made you uncomfortable*; instead say *It seems that the way this is presented raised an issue with which you are uncomfortable*. Use the passive voice. Do not make this a personal fight as in *You are insinuating I doctored the data* or admit responsibility for the hurt as in *I am sorry I made you feel this way*. Make this a fact against fact issue, not a person against person issue. If the person says *I am very upset*, rephrase by toning down the word *upset* with *uncomfortable* or another word that expresses a milder feeling such as *affected by*. Indeed, all offensive words in the question are to be removed during rephrasing. You thereby demonstrate your qualities as a respectful and caring host — and you buy yourself extra time to think of an answer.

2) It is always necessary to wait until the questioner acknowledges and agrees with your rephrasing prior to responding. There is no need for you to smile during the rephrasing process: it could be misinterpreted as not taking the questioner seriously.

3) When answering, find a way to agree with some of the statements (not all) used by the questioner, therefore giving your guest a face-saving exit. For example, if the questioner says that a similar experiment led to very different results in his research lab, do not disagree. It would be equivalent to calling him a liar! If he says so, then it must be as he says. You are not here to dispute the quality of someone's work; you are here to defend the quality of yours. Here is a possible answer: *Sir, I do not for one moment question the results you obtained in your lab. There may be many reasons for the differences between what you have observed and what we have observed: they could be explained by the type of equipment, the data sampling method, the population, the cell-lines, and even leaching plastic ... There are so many factors that interplay and contribute to a result.*

4) When you have accepted the questioner's assertions, defend your results and conclusions ON SOLID SCIENTIFIC GROUNDS, not on feelings. Restate what you have observed, but this time mention the quality and care taken in your methodology. If other papers agree with some of the

results you have found, indicate that also. Do not appear nervous. Remain calm, self-assured, and authoritative at all times.

5) When you have finished responding, do not look back to the hostile questioner; look in a very different direction, so as not to appear interested in continuing that line of questioning. Of course, do not ask whether you answered the question to the satisfaction of the questioner. Hostile questioners often behave like serial Q&A killers, they may return for a fight. If such a situation occurs, do not rephrase. Simply say that you already answered the question and have no further comment on that topic.

Do not rely on the chairperson to intervene unless the words used in the question or comment are offensive, sexist, or insulting. The chairperson may be happy to see a little bit of action in what seems a long uneventful day. The audience may also be interested to see how you are going to get out of a difficult situation. If you do it properly, and the hostility continues, the audience will actually lose its neutrality and some may start to openly disagree with the questioner. However, do not let the discussion degenerate until people in the audience are arguing with others while you silently wait until the disruption ends. Seize the mike and speak loudly to say something like the following: *Here are the concerns I hear from both sides*, then say that it is best to leave this open debate alone for the time being, and ask for other questions.

Typical Questions from Specific Groups

Questions from Scientists

In this chapter, we examined 14 types of questions that may be asked during a public scientific presentation: the come-onto-my-turf question, the boundary question, the have-you-tried question, the why-didn't-you question, the what-if question, the poke-a-hole question, the IP question, the senseless question, the teach-me-the-basics question, the unintelligible question, the crystal ball question, the time-consuming question, the you've-got-me question, and the hostile question. However, some questions are more general, such as the following:

Why have you chosen this method as opposed to others?
Why are you doing this research? What is its expected impact?

What in your contribution can be used concretely and immediately?
Why have you decided to tackle this problem?
What is novel in your presentation?
What do you intend to do next?

Questions from Board Members

Board members are not your typical audience. They sit through all presentations, including those outside of their direct field of expertise. They probably don't have enough technical background to fully appreciate your scientific contribution. Board members are usually chosen because of the breadth of their knowledge of both science and people. They are aware of what other labs in the world are doing and have a developed sense of the applicability and potential of scientific findings. They will want to explore further the resemblance to related works, and question the uniqueness of the research. They will also assess the state of readiness of a contribution for technology transfer and offer suggestions to solve research problems. They will question the appropriateness of a research when other opportunities seem to have a narrower time-window, or when the research does not seem to lead to paradigm-shifting outcomes. Because board members are not always knowledgeable in your field, it is best to answer them simply, without great technical details. However, because they have a high academic level, they are able to learn "on the spot" and probe for details that are more technical. Therefore, be prepared to "fill them in" at a higher than expected level. It never pays to underestimate a board member.

Questions from Project Managers

Critical path questions:
> Is this a priority? Aren't there tasks that are more urgent?
> What is the time schedule?
> What are the contingency plans?
> How are you going to allocate your resources?

Worthiness questions:
> Is the research leading to a paradigm shift?
> Have any other people expressed interest in this?
> What are your prospects for a grant?

Feasibility questions:

> Do you have people with the right skills, and if not, can you get these people?
> How much will it cost?
> How much time is required to do this project?
> Have you identified partners?
> Do you have a working prototype?

Questions from Investors

Questions on the value of the technology/discovery:

> How unique is this, and who else knows about this?
> Are there alternative ways to do the same things?
> Do you have patents?
> Do you have an unfair advantage?
> What are the expected markets and their relative sizes?
> What is the time to market?
> How sustainable is this?
> Are there any interdependencies with other technologies?
> At what stage of development is this: idea, prototype, deployed, in use?

Questions on the value of the research team members:

> Who contributed to the breakthrough thinking?
> What is the business acumen of the team?
> What is their background? Any scholarship, award, citations?
> Do people have a "can-do" attitude?

Techniques for Fast Answer Support

You now know that it is vital to keep technical slides after the conclusion slide of your presentation, just in case the experts in the room ask for more precision, or more information. You also know that sometimes, the audience asks you to return to one of your slides. Your goal is to immediately bring to the screen the slide requested, or the slide that will help you with the expert question, so that no precious Q&A time is wasted in time-consuming slide retrieval manipulations.

Presentation software, such as PowerPoint 2004 or Keynote II, provides ways to do this. Some of these ways can be disruptive (for example, the underlined text link or the numbered slide) or time consuming (the nested navigation sub-menus). It is important for you to know the best methods to rapidly, transparently, and directly access any slide in your presentation.

Secret Hyperlinks in PowerPoint 2003

First, let us examine a direct, transparent, and rapid way to go to a particular slide: the hyperlink. Hyperlinks are easy to create. Any word, line of text, image, or shape can be a hyperlink — a springboard to jump directly to the linked slide. Hyperlinks should not be visible. They are your secret. Only you know the secret door to the tunnel that will lead you to another room in your castle of slides. Why should they be a secret? Simply because people recognise a text hyperlink when they see one: the text is underlined or changes colour when clicked. When people identify a hyperlink and do not see you click on it, they wonder why you are hiding things from them. Non-text hyperlinked objects will not betray your secret. So do not select text and hyperlink it to another slide. If you do wish to use text as a hyperlink, but an invisible one, there are three ways to do this.

1) Type the text, pull down the *Insert* menu, select *Text Box* and insert the text box over your text (make sure there is text under your text box!). Text boxes are transparent. Handles should be visible at each corner of the rectangular text box (if they are not visible, insert the text box again, and do not click anywhere else on the background afterwards). Pull down the *Insert* menu again, select *Hyperlink*, and then click on the *Document* tab, and click on the *locate* button; then select the slide you want to jump to when the link is activated and click on the *OK* button. Go into slide show mode. Note that the text has no underline, which is what you want. As you move the mouse over the text, the pointer changes into a finger and clicking the text will activate the link and take you to the chosen slide.

2) Draw a rectangle (using *insert picture/autoshape*). Right click on it, and select *format autoshape*, make its colour transparent (*no colour*) and select *no line* in the line colour menu. Then, as the handles are still showing,

type in your text; the rest is the same as above: pull down the insert menu again, select *Hyperlink* … You can always change the size of the rectangle or the size of the text later.

3) Type your text. In the slide show menu, select *Action Buttons*, and then *Custom*. Your cursor changes into a crosshair. Draw a shape above your text. Then, in the dialog box that appears, hyperlink the button to the linked slide. As you click *OK* to close the dialog box, you return to your button. Right click on it. You will open the *format autoshape* dialog box. Select *No fill* in the fill colour menu, and *no line* in the line colour menu. You now have a transparent button on top of your text.

Secret Hyperlinks in Keynote

Actually, in Keynote, it is easy to remove the underline from text hyperlinks. You could set the keynote preferences (under the auto-correction tab) and deselect the option *Underline text hyperlinks on creation*. Alternatively, you could simply select the underlined hyperlinked text, then go to the Font dialogue box, and select *none* in the pop-up menu found under the triangle to the right of the underlined T (left of dialog box).

All other objects can also be hyperlinked, as in PowerPoint. The techniques for creating an invisible button in Keynote have been explained in the PowerPoint/Keynote chapter.

The Number Trick

People really familiar with PowerPoint and Keynote know that typing the number of a slide followed by enter (or return key) while in slide show mode will instantly bring that slide into view. Therefore, it is always important for you (not the audience) to know what is on a slide, and to know the number of that slide in your slide show. One way to do that is to take note of the title and number of each slide on a small card you carry in your pocket. Others print nine miniature slides per page (the maximum allowed by PowerPoint); Keynote allows only six. They keep these pages next to the computer, ready to use to identify the number of any slide. Using Photoshop or other graphic applications, you can put more than nine slides per page (save each slide as a png file, drag and drop it into the application, then reduce its size).

Build Avoidance Techniques

Those of you who regularly use builds (Keynote), or custom animations (PowerPoint), know that coming back to a slide takes you to the beginning of that slide, right before the effects/builds layers, if you have any. If what you need to show is on the last effect/build, you will have to click through each effect in turn before arriving there. This is cumbersome. There is another way of course, but a distracting one. You could, for example, go to the slide that follows that slide, and then click the left arrow key to move to the last effect of the previous slide. There are better ways, however.

1) Duplicate a slide, remove its effects, and hide it (PowerPoint only). If your build effects do not make use of the "exit" effect provided by PowerPoint's custom animations, the following applies. Duplicate the slide that has numerous effects. Place it immediately after the original slide. Remove all the effects on the new slide, right click on it in the slide sorter view, and select *Hide Slide*. When you play your slide show, that hidden slide is skipped and does not appear. When you have to answer a question with that slide, just type its number, and it will appear without you having to go through all the effects.

2) Replace effects by slide transitions (PowerPoint and Keynote).

 Watch the "Effects without effects" video on the DVD

Imagine you have three effects/builds on one slide. Instead of having one slide, you could have four slides. Each successive slide would be a duplicate of the one that precedes it, plus one additional effect. Every time you click, you move to the next slide instead of bringing the next layer of animation. The result is the same, but you now have four slides instead of one. You could then type in the number of the slide that helps you answer the question. Doing so, however, increases the number of slides, so you may have a thick wad of paper filled with your miniature slides!

Conlusion

At the end of this book, you may feel overwhelmed by the sheer number of techniques to master, and the quantity of small details worthy of your attention. As Victor Hugo says in "*Les Misérables*", there are no small details. All are necessary.

> *"And yet, these details, that people wrongly qualify as small — there are neither small facts in humanity, nor small leaves in vegetation — are useful. For it is from the physiognomy of years that centuries take form." Les Misérables. Chapter 1, last paragraph.*

The sweating out of hundreds of small details forges the considerable talent of a presenter. This takes effort. It takes the perseverance of the forger who hammers a red iron to its perfect shape, repeatedly, blow by blow. Rehearsals after rehearsals, your presentation improves. Each time you consider the audience, your presentation improves even more. Every new idea convincingly put forward is remembered alongside your face. You and your science, the host and the scientist, are seen as one.

Throughout the book, you have been invited to explore the companion DVD. It contains a wealth of information, audio and video: useful get-to-know PowerPoint and Keynote techniques to improve your presentation, and podcasts sharing the views of three scientists on scientific presentations: Dr. Alastair Curry, Dr. Mark Sinclair, and Dr. Juzar Motiwalla. They speak as presenters, but also as members of the audience, and even as session chair. These resources, plus the numerous websites mentioned in these pages, complement the book. More resources are provided for your enjoyment on the author's websites and blogs: **www.scientific-writing.com** and **www.scientific-presentations.com**

It has been my pleasure to accompany you part of the way on your life journey, a life where each successful scientific presentation is a milestone, a springboard to a more fruitful career. Presenting is fun, even though it is work. Let it be heart pumping, not heart wrenching. And when you face your audience, reward it with that gorgeous smile of yours, the one your friends enjoy. Here is mine.

APPENDIX

The Podcasts

Podcast Guests

Dr. Alastair Curry. British geomorphologist. Background in Paraglacial landsystems and Holocene landscape evolution. BA, MA, University of Oxford; PhD University of St. Andrews 1995–1998. Successively, lecturer and senior lecturer at the University of Hertfordshire, 1999–2004; and now volunteer lecturer in Physical Geography at the Royal University of Phnom Penh, Cambodia, where he teaches, trains lecturers and is helping to develop teaching material in Khmer. He attended countless presentations and presented at a number of national-level conferences.

Dr. Mark C. Sinclair. He started his career as a systems designer at GEC Plessy Telecommunications (GPT) in Liverpool. He was then appointed as lecturer in Electronic Systems Engineering at the University of Essex, Colchester. Now with OMF in Cambodia, he was initially seconded to the Royal University of Phnom Penh as Professor of Computer Science, but is currently Professor of IT at the National Polytechnic Institute of Cambodia (NPIC). His research interests include the design of very wide area networks, using a variety of naturally inspired heuristic algorithms, including evolutionary algorithms, particle swarm optimisation, and ant algorithms. He has organised workshops and presented at many national and international conferences.

Dr. Juzar Motiwalla. He is Professor in the practice of Entrepreneurship at the National University of Singapore (School of Computing) and sits on the board of several companies in the technology space that have global footprints. Prior to that, he was partner at Green Dot Capital, a Venture Capital firm where he was responsible for global investments, and CEO of Kent Ridge

Digital Labs (KRDL), a 350 person high-tech laboratory. KRDL incubated 15 start-up companies in Asia and Silicon Valley. Early in his career, after a PhD at the University of Wisconsin–Madison in 1977, he joined the National University of Singapore's Department of Business Administration and in 1983 was named Director of the University's Institute of Systems Science where he led the development of several large international ventures with U.S. computer companies, in particular Apple and IBM.

Podcast Topics

Chapter 1

- Keeping to time
- What are the benefits (of presenting)?

Chapter 3

- Not so expert audience with distracting laptops
- What does the audience remember?
- PowerPoint and Shakespeare
- Dealing with accent

Chapter 4

- Stating limitations

Chapter 5

- Core competitive advantage

Chapter 6

- Pearls of presenter wisdom
- Three audience irritants

Chapter 8

- Presenter mistakes

Chapter 9

- David Peeble's argument
- Dealing with accent
- Personal credibility

Chapter 10

- The chairperson and the presenter
- Attitude of audience toward presenter during Q&A

The PowerPoint and Keynote Videos

Chapter 5

- Map slide
- Moving through a large document

Chapter 6

- Map slide

Chapter 7

- Presentation remotes
- Microphones
- Scale groups of objects/images
- "B" Key
- Animations

Chapter 10

- Effects without effects

Index

Audience
Attention, 15, 49, **124–135**, **194–196**
Expectations, 14, 27, 31–32, 37, **52–64**, 72, 74, 76, 85–86, 92, 127, 195, 213
Expertise, 13
Memory recall, 13, 15, **43–45**, 57, 78–80, 92–93
View of the presenter, 13, 176–177

Content selection criteria
Applicability, 33–34, 62
Audience, 14
Believability, 56
Novelty, 15, 33–34
Time to explain, 33–34
Title, 14, 22, 33–34

Exercises
Assess believability, 62
Assess usefulness of talk to audience, 64
Assess your degree of conviction, 144
Check contents of title slide, 32–33
Determine audience from conference type, 54–56
Determine audience from title, 27–31
Select slide contents, 34–36

How to
Check legibility, 112–113
Choose a good font, 101–108
Convince, 16, 20–21, 36, **56–62**, 74, 84, 88, 133, **144–147**, 185–186, **196–199**
Emphasise, 89, **128–131**, 168, 175, 194, 198–199
Gesture, 178, 208, 209, 219
Make use of the chairperson, 169, 172, 187, **204–205**, 209–210, 219–221, 233
Speak clearly, 187–193
Use a "B" key (or black slide) effectively, 82, 88, 93, 125, 152–153, 173, 175
Write a good slide heading, 128, 147

PowerPoint/Keynote Techniques
Adjust image brightness and contrast, 122, 168
Animate bullets one at a time, 167
Arrange slide order, 90–91

249

Build a delayed effect, 131
Create a text hyperlink without text highlight, 240
Create a transparent button, 140
Create an image from a slide, 167
Create an image from grouped items, 176
Create effects without effects, 242
Determine image size once dropped in slide, 119–121
Disable automatic text formating, 105
Disable word wrapping, 106
Establish number of points per cm, 112
Export slide as png file, 114
Hyperlink a button to a slide or another presentation, 165
Hyperlink an image to a slide, remove a hyperlink, 165
Increase size of slides in slide sorter or light table, 139
Insert and play a movie on a slide, 166
Jump directly to a slide, 241
Make a black slide, 167
Remove the Master background from a slide, 167
Resize image to fit slide format, 121–122
Simulate "B" key with black slide, 168
Skip a slide or jump to a hidden slide, 242
Use motion path or action builds, 168

Presentation
Mistakes, 40–42, 77, 155
Purpose, 13, 43, 93
Room, 99–100
Start, 69, 71, 81, 184
Title, 14

Presentation tools
Blank card, 214–215, 221–222, 241
Business card, 64, 180, 217
Laser pointer, 155–157
Lights-audio, 164
Microphones, 158–162
Presentation remote, 154–158
Printout of miniature slides, 241

Presenter
Accent, 47–49, 72, 187, **191–193, 232–233**
Adrenaline, 177, 183–184
Behaviour, 14, 51, 54, 187, 189, 210, 227
Breathing, 152, 183–191
Clothing, 160–161, 177–179, 186–187
Co-host, 77, 94, **174–175**
Fear, 51, 169, 176–177, **182–184**, 190
Food/Drink, 183–184, 189
Gestures and body language, 175, 178, 185, 198, 221, 233
Host, 149, **171–181**
Name, 71
Personality, 214
Position on stage, 151–154

Posture, 152, 177, 184, 189, 190

Replacement, 72, 234

Role, 2, 170

Stress, 184

Time control, 15, 16, 41, 66, 77, 137–143, 180, 188, 204

Presenter qualities

Attentive, 49, 54, 173

Available, 63–64, 180

Clear, 41, 62, 78, 195

Confident and expert, 51, 59–60, 146–147, **182–186**, 196, 222

Credibility/Authority, 19, 51, 59, **196–198**

Discrete, 179–180

Intellectual honesty, 60, 145–146, 180, 196–197, 216

Interesting/animated, 175, 195

Networker, 63–64, 180, 209, 215, 217, 227, 226

Positively charged, 50, 185

Prepared/rehearsed, 172–173

Respectful and patient, 51, 69, 181

Welcoming, 50, 172–173

Q&A

No question situation, 94, 209, 215–216

Pre-empting questions, 94, 145–147

Process, 204–217

Role, 15, 16, 19, 180–181

Slide support for …, 92–94, 229, **240–242**

Question

Answering, 215–217

Asking style, 218–222

Preparation, 205, 207, 222

Rephrasing, 211–213

Types, 9, 13, 14, 60, 62–63, **222–239**

Rehearsal

How to, 42, 184–185, 197, 207

Role, 16, 34, 41, 48, 158, 184, 196

Towards less is more, 86, 95–96, 189

Slide

Design, 124, 148

Direct access to …, 140–141

Duration, 15, 54

Heading, 90, 95, 128

Organisation, 42, 80–81, 84–85, 90–91, 142

Transition, 18, 42, 54, 82, 85, **87–90**, 93, 195

Slide qualities

At the right place in the story, 84

Convincing, 57–59, 75, 133, 144

Fast to comprehend, 41, 54

Legible, 5, 97–124, 135

Linked to title, 25, 26

Self-contained (no dependencies), 34,

Uncrowded (see also *less is more*), 84

Visual, 34

Slide type
  First slide, 18
  Title slide, 14, 17, 32–33, 40, 44, **68–72**, 75, 78, 92
  Acknowledgment, 32, 71, 92
  Hook slide, 70, **73–79**, 81–84, 87, 90, 92, 139, 141, 185
  Map/Outline slide, 18, **79–82**, 89–90, 132, **141–142**
  Story slide, 18, 24–27, 42, 74, 78, **83–91**, 93–94, 147
  Conclusion/Summary slide, 42, 82, **91–96**, 124, 139, 141–142, 185, 208
  Supplementary slide, 34, 36, 41, 95, 122, 222, 227

Visual
  Alignment, 123–124
  Animation/Effect, 33, 36, 40, 78, 89–90, **130–135**, 242
  Design, 34–35, 45
  Hook, 76
  Presentation, 44
  Rework, 36, 121–122
  Size, 117

Vladimir stories
  Blinded, 6–7
  Click to add title, 67
  Compression exercise, 3–4
  Dr. Sorpong's masterful preparation, 24–27
  Dr. Sorpong on the grill, 203–204
  File://localhost/Users/Vladimir/Desktop/Vladprezo.ppt, 135–137
  Maui Hotel, 46
  Mrs. Toldoff receives her guests, 170–171
  Vladimir backstage exit, v, vi
  Vladimir on the grill, 201–203